Lecture Notes in Economics and Mathematical Systems

463

Springer-Verlag Berlin Heidelberg GmbH

Gilbert Abraham-Frois (Ed.)

Non-Linear Dynamics
and Endogenous Cycles

 Springer

Editor

Prof. Gilbert Abraham-Frois
University Paris X Nanterre
Centre National de la Recherche Scientifique
200, Avenue de la Republique
Batiment G
F-92001 Nanterre Cedex, France

Cataloging-in-Publication Data applied for

Die Deutsche Bibliothek - CIP-Einheitsaufnahme

Abraham-Frois, Gilbert:
Non-linear dynamics and endogenous cycles / Gilbert Abraham-
Frois. - Berlin ; Heidelberg ; New York ; Barcelona ; Budapest ;
Hong Kong ; London ; Milan ; Paris ; Santa Clara ; Singapore ;
Tokyo : Springer, 1998
(Lecture notes in economics and mathematical systems ; 463)
ISBN 978-3-540-64321-0   ISBN 978-3-642-58901-0 (eBook)
DOI 10.1007/978-3-642-58901-0

© Springer-Verlag Berlin Heidelberg 1998
Originally published by Springer-Verlag Berlin Heidelberg in 1998

Typesetting: Camera ready by author
SPIN: 10649636   43/3142-543210 - Printed on acid-free paper

Table of Contents

Part 4: Methodological Issues

Introduction

This book is the result of a workshop organized by MODEM, research unit supported simultaneously by French "Centre National de la Recherche Scientifique" (C.N.R.S) and University of Paris-X-Nanterre on May 23-24th 1996.

When dealing with "*Non-linear dynamics and endogeneous cycles*", it seems quite difficult, and rather contradictory to give a clear separation between cycles and growth. However, it may be used as a useful starting point for presentation of two kinds of papers, a first lot one more concerned with " economic growth models", a second mainly with " business cycles" . Moreover, it has appeared that other papers could be characterized briefly as dealing with "keynesian models", forming consequently a third part of this volume, the fourth part being centered about methodological issues, concerning predictability problems and differences between stochastic and deterministic systems.

Considerable work has been done on chaotic dynamics in the field of economic growth and dynamic macroeconomics. The study of chaotic dynamics in economic growth has a root in a paper dating back from 1982 by Richard Day. It is a great pleasure to open the first part of this volume with a contribution due to R.Day, Z.Wang and M.Zhang dealing with "*Infrastructure in an Adaptative Economizing Model of Economic Growth*". This study presents an adaptative economizing model of economic growth based on boundedly rational agents that incorporates infrastructure in terms of physical and human capital and a utility function based on a lexicographic preference ordering of present and future potential consumption. The introduction of infrastructure induces a restriction on capital labor substitution which, when combined with imperfect foresight, permits factors to be used inefficiently. In such cases the marginal products of capital and labor used in production are always positive but the marginal product of the *total* work force or that of the *total* capital stock is negative. Because of the opportunity cost of resources used in infrastructure, the usual competitive equilibrium conditions are modified. The authors show that capital accumulation trajectories are both generically asymptotically stable and generically unstable, converging to a steady state or fluctuating around one depending on the weight given by a given generation to its heirs.

G.Dufrenot's contribution (" *Neo-classical Growth and Complex Dynam-*

ics: A Note on Day's (1982) model") considers new aspects of Day's original (1982) contribution. Two aspects of his paper are under discussion here. Day's contribution may be viewed as a prototype of a model that can exhibit chaotic dynamics with an ad hoc modification of the standard text-book approach to neo-classical growth theory. The first aim of this paper to provide a plausible explanation which can be used to justify the presence of the "pollution effect": this effect can be reinterpreted within the framework of a disaggregate economy, as a consequence of heterogeneity amongst entrepreneurs, specially in their expectations behaviour. Such reformulation of Day's seminal model suggest that volatile expectations may be a cause of growth irregularity even in a one-sector economy, when commodity markets are imperfect. Another aspect of complex dynamics is under discussion in Dufrenot's paper. By using bifurcation analysis, it is shown that solutions to non-linear difference equations that produce irregular fluctuations are highly sensitive to a variation of the parameters. This implies that even when conditions for stable growth are present, transient chaos may be observed; consequently, the economy may be quite long to converge towards a sequence of capital accumulation which is monotone over time.

Three papers centered on business cycles analysis constitute the second part of this volume. The first one, due to R.Franke, appeals to the concept of hysteresis which has become widespread in explanations of a variety of economic phenomena. The two other ones are concerned with overlapping generation models.

R.Franke's paper (*"Hysteresis Arising From Individual Inertia and Behavioural Heterogeneity"*) emphasizes a natural sort of individual inertia and behavioural heterogeneity as a source of hysteresis in aggregate variables. It sets up a deterministic dynamical model which, however, cannot be reduced to a standard system of equations. Studying perturbations of a growth equilibrium, the economy may converge to different new steady states or approach different periodic growth cycles. The hysteresis effect in the first case turn out to be asymmetric. A special feature is that the economy might enter into persistent fluctuations in response to a medium-sized shock whereas it finds to a (new or old) equilibrium if a small and also if a sufficiently large shock occurs. In a scenario of repeated stochastic shocks, it is indicated that a random walk hypothesis is unlikely to be a useful approximation to describe the evolution of the (equilibrium) growth rates.

Two other papers are concerned with cycles in overlapping generation models. J.M.Grandmont (1985) was among the first to study the possibility of self-sustaining cycles in the overlapping generations models. In his paper, the origin of these cycles is the conflict between the substitution effect and the income effect due to the variations of relative prices. Grandmont's paper has been criticized on empirical basis: income effects in the model are two large compared to their values in real life. G.Dufrenot and L.Mathieu's paper ("*Non linear dynamics and utility functions in over-lapping generations model*") provide some theoretical arguments that may explain why cyclical and complex paths cannot be ruled out. Following an early suggestion by Drèze and Modigliani, and using a methodology initiated by Sato, they show a close connection between the hypotheses made on the Slutzky equations corresponding to excess demand functions and classes of utility functions from which cycles and complex dynamics emerge.

In the same frame of analysis (OLG models), M.Botomazava and V.Touzé examine "*Pay-as-You-Go System under Permanent Business Cycle*". Relying on Reichlin's (1986) model, they give conditions for the existence and stability of cycles when a pay-as-you-go system is introduced. To obtain a permanent cycle, a condition is crucial. Nevertheless, this cycle around the steady state level can be interpreted as two intergenerational inequality criteria. Some computing results are specially interesting: when the payroll tax rate rises, the lowest welfare level (Rawlsian criteria) and the intergenerational inequality (egalitarian criteria) can grow up.

Two papers centered around "keynesian models" constitute the third part of the book. R.Goodwin (1951) was probably the first to introduce "relaxation cycles" in macro-economics. Hicks's analysis is indeed anterior (1950), both authors introducing "non-linear accelerator" in their analysis; but Goodwin's originality is obvious since he has used relaxation cycle and Liénard-van der Pol equations. Further work has shown that in many cases relaxation cycle can be considered as a stable "attractor", a rather peculiar "limit-cycle", named in some cases "limiting limit cycle" (Chiarella 1990). Moreover, it is well known that cubic functions can give rise to chaotic dynamics. In their paper ("*Relaxation Cycle, Chaotic Dynamics and Limit Cycle: a Model with Keynesian 'Flavour'*"), G.Abraham-Frois and E.Berrebi, using previous work by Puu (1991) and themselves (1995) study the relationship between relaxation cycle and chaotic dynamics. For certain values of the parameters, there is simultaneous appearance of chaotic dynamics and relaxation cycle; moreover, relaxation cycles are part of a cubic which constitute (with its immediate neighborhood) an attraction basin; chaotic part of the cubic may be inside or outside the relaxation cycle, which appears as an attractor for the cubic, itself an attractor for the model and its neighbourhood. In some other cases, one observes relaxation cycles with "thick frontiers" and even dislocation of the relaxation cycles. At last, one can observe formal similarities between "classical" (Feigenbaum) bifurcation diagram and cubic.

R.Arena and A.Raybaut's paper gives us an interpretation of "*Credit and Financial Markets in Keynes' conception of Endogeneous Business Cycles*". Of course, it is well known that J.M.Keynes never developed a proper and systematic theory of the business cycle. However, the *Treatise on Money* and the *General Theory* include, according to R.A & A.R substantial indications on which it is possible to rely on order to explain the macroeconomic emergence and persistance of economic fluctuations. The only obstacle lies in the apparent contradictory elements which are present in Keynes's works; according to to the post-Wicksellian viewpoint of the *Treatise*, the emphasis is put on the divergence between the natural and the market rates of interest and on its effects upon the demand for credit, while the *General Theory* stresses the role of human psychology (including on the stock exchange) in the determination of investment decisions. The paper shows how it is possible to build a single framework allowing to cope with both components of the contradictions and, therefore, to explain why it was only apparent.

Last part of this volume is concerned with methodological issues, more specifically with the difference between determinist and stochastic systems, which seems to meet some difficulties here. The conclusion of A.Medio's paper (*"Business Cycles, Chaos and Predictability"*) is *"that the implications for economists are puzzling"*. The author consider the case in which a model of optimal growth gives rise to a dynamic equation of a "logistic" type with chaotic parameter. When the system is Bernoulli, agents face sequences of values which have the same probabilistic structure as random processes. Consequently, for a certain class of concrete dynamical systems, the possibility exists of repressenting them either as deterministic systems (plus perhaps some random disturbances) or as stochastic processes. And the conclusion is quite pessimistic about any hope of providing a general test for distinguishing deterministic chaos and "true" randomness.

The last paper by G.Abraham-Frois, S.Lardic and V.Mignon (*"Long-term Memory and Chaos"*) has a a similar preoccupation with a different point of departure. It is well known that the Hurst exponent H, which allows to detect the presence of long-term memory, is linked to the fractal dimension D of the underlying process by the relation D = 2 - H. Thus, it may be tempting to set a link between long-term memory and chaotic processes by the mean of D. Moreover, another relation does exist with ARFIMA (Auto-Regressive Fractionally Integrated Moving Average) processes, which are a generalization of standard ARIMA (p,d,q) processes with d fractional (and no more an integer): this particular class of discrete time models, is linked to fractional Brownian motion defined by Mandelbrot and van Ness and characterized (also) by Hurst exponent. But the estimation of the value of fractional differencing parameters is far from being conclusive. At last, one comes to same kind of perplexity as in precedent paper since same process might be considered either as deterministic or as stochastic.

This brief introduction would not be complete without mentioning, and thanking, other contributors and discutants of this wokshop, especially A. d'Autume (MAD, Paris I), Ch. Bidard (MODEM, Paris-X), J. Blot

(CERSEM, Paris I), P. Cartigny (GREQAM, Aix-Marseille II), G. Cazzavillan (Venice), R. Courbis (MODEM, Paris-X), M. Currie (Manchester), J.P. Drugeon (MAD, Paris I), A. Goergen (MODEM, ENS-Cachan), J.M. Grandmont (CEPREMAP, Paris), I. Kubin (Munich), T. Lloyd-Braga (Lisbon), P. Pintus (EHESS & CEPREMAP, Paris), W. Nisima-Calmel (MODEM, Paris-X), G. Rotillon (MODEM, Paris-X), D. Sands (Orsay), R. Topol (OFCE, Paris), R. de Vilder (Amsterdam & CEPREMAP), A. Venditti (GREQAM, Aix-Marseille II). Financial assistance was provided by CNRS and University of Paris-X-Nanterre.

Special thanks to V.Touzé for (precious) editorial assistance.

G.ABRAHAM-FROIS

Part 1

Economic Growth Models

Infrastructure in an Adaptative Economizing Model of Economic Growth

Richard H. DAY, Zhigang WANG and Min ZHANG

1 Introduction

This study presents an adaptive economizing model of economic growth based on boundedly rational agents that incorporates infrastructure in terms of physical and human capital and a utility function based on a lexicographic preference ordering of present and future potential consumption. We show that capital accumulation trajectories are both generically asymptotically stable and generically unstable, converging to a steady state or fluctuating around one depending on the weight given by a given generation to its heirs.

Our model is a descendant of the L–D–L growth model originally set forth by Leontief (1965) and developed by Day (1967), Lin (1987), and Day and Lin (1991). Although this approach is closely related to the temporary equilibrium analysis explored in macroeconomic settings by Grandmont and Laroque (1973), (1986), (1990) it is not equivalent even in the case of a single generation. This is because the introduction of infrastructure induces a restriction on capital labor substitution which, when combined with imperfect foresight, permits factors to be used inefficiently. In such cases the marginal products of capital and labor used in production are always positive but the marginal product of the *total* work force or that of the *total* capital stock is negative. Also, because of the opportunity cost of resources used in infrastructure, the usual competitive equilibrium distribution conditions are modified.

Section 1 describes preferences. Section 2 reviews the RFS production function, and section 3 derives the consumption/savings strategy when preferences have a specific lexicographic character and adaptive expectations are naive. Section 4 considers the existence and the stability/instability properties of capital accumulation paths.

2 Preferences

Distinguish between historical time, represented by a *subscript* and *planning time*, represented by a *superscript*, which represents the anticipated future. For example, if x were a decision variable, x_t^i would be the value to be acted on i periods in the future but determined "now" where "now" is at historical time t. Then x_t^0 is the value of the decision variable planned for the current period.

Let x_t be the value actually acted on in the current period. Then we also assume the temporary equilibrium condition, $x_t = x_t^0$, that is, the act planned for the current period is carried out. Plans for the future, however, are not always carried out so that it could (and usually will) happen that $x_t^i \neq x_{t+i}$.

As a further convention, when analyzing the planning problem of the foreword looking agent, we treat x^i as the decision variable to be determined and x_t^i the value that is determined for the historical time t. In this paper we only consider agents who plan ahead a single generation. Let c^0 be the level of current consumption to be determined "now" and c^1 be the potential future standard of living in perpetuity, also to be calculated "now." Since c_t^0 will be temporary equilibrium in our model, we can drop the superscript in c^0, denote by c the current level of consumption to be determined and by c_t the level actually achieved.

It is convenient to think of the economy as a sequence of overlapping generations. If x_t is the number of households, then the number of children per family is $2(1+n)x_t$ and the number of households in the next generation is $x_{t+1} = (1 + n)x_t$. Adults of a given generation are manager–worker–owners to whom all proceeds of production are distributed. They determine consumption and savings for themselves and their children. The savings are invested and the capital stock that results, allowing for depreciation in the meantime, is the bequest which constitutes the endowment for the next generation of adults. Suppose the adults explicitly consider the trade–off between their own consumption and the standard of living which their heirs can potentially enjoy and determine their behavior accordingly. The adults of the next period then repeat the same economizing decision but on the basis of the capital stock inherited from their predecessors.

With these conventions the preferences of the current generation of adults are represented by a utility function,

$$
u\left(c, c^1\right) = \begin{cases} v\left(c\right) & , 0 \leq c, 0 \leq c^1 \\ v\left(\bar{c}\right) + \psi v\left(c^1\right) & , \bar{c} \leq c, 0 \leq c^1 \leq \bar{c} \\ (1 + \psi)\, v\left(\bar{c} + w\left(c - \bar{c}\right) + \psi v\left(c^1 - \bar{c}\right)\right) & , \bar{c} \leq c, \bar{c} \leq c^1 \end{cases}
$$

(1)

where $v(\cdot)$ and $w(\cdot)$ are twice differentiable, $v' > 0$, $w' > 0$, $v'' < 0$, $w'' < 0$ and $v(0) = w(0) = 0$.[1]

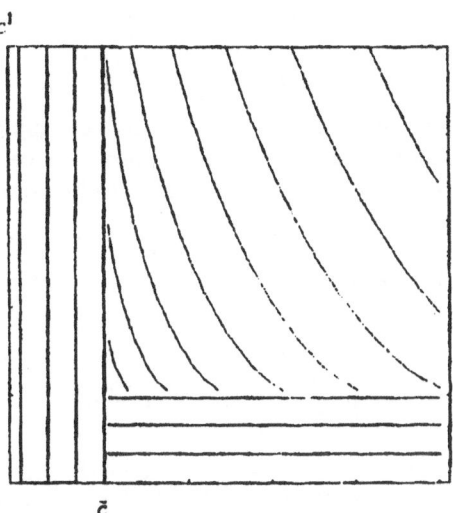

Figure 1: The $\mathcal{L}^{**}$ Preferences Between Current Consumption and the Potential Future Standard of Living

This utility function means that there is a threshold, $\bar{c}$, short of which no utility accrues from considering future consumption. Above that threshold, the sustainable standard of living that can potentially be enjoyed by the next generation is given priority over additional current consumption. Then a second threshold exists (also $\bar{c}$) in the level of anticipated (future) consumption. If these thresholds are exceeded, current consumption and the anticipated future sustainable standard of living both increase current utility. It is well known that the preferences thus described cannot be represented by a continuous utility function. However, the most preferred choices do depend continuously on income.[2]

---

[1]If both c^1 and c exceed $\bar{c}$ and if each future generation actually maintained the capital stock it received, then $\psi\left[v(\bar{c}) + w(c^1 - \bar{c})\right]$ would be equivalent to the discounted sum of the utilities of an anticipated constant stream of consumption c^1. Thus,

$$\alpha[v(\bar{c}) + w(c^1 - \bar{c})] + \alpha^2[v(\bar{c}) + w(c^1 - \bar{c})] + \ldots = \frac{\alpha}{1-\alpha}[v(\bar{c}) + w(c^1 - \bar{c})]$$
$$= \psi[v(\bar{c}) + w(c^1 - \bar{c})]$$

where α is the time preference with $0 < \alpha < 1$ and $\psi = \alpha/(1 - \alpha)$. As noted above, however, in general $c_t^1 \neq c_{t+1}$ so that this equivalence can hold exposte only at a stationary state. See above.

3 Technology

Consider the conventional production function

$$Y = BF(K, L) \tag{2}$$

where Y is aggregate output, L is the available supply of labor, K is production capital, and B represents the "neutral" level of technology. It is assumed that $F(\cdot)$ is concave, twice differentiable and homogeneous of the first degree, that

$$F(0, L) = F(K, 0) = 0 \qquad \text{for all} \qquad L, K \geq 0,$$

and that

$$\lim_{K \to 0} \frac{\partial F}{\partial K} = \infty = \lim_{L \to 0} \frac{\partial F}{\partial L}.$$

In the spirit of Boserup (1981) and North (1981) infrastructure is taken account of in the following way. Total capital, Z, consists of the component, K, employed in market production, and the component, $\bar{K}$, employed in the socioeconomic infrastructure,

$$Z = \bar{K} + K \tag{3}$$

The total work force, x, consists of the *labor force* employed directly in market production, L, and the *infrastructural work force*, $\bar{L}$. That is,

$$x = \bar{L} + L \tag{4}$$

We assume that one adult equivalent is used in activities internal to the family and one adult equivalent is allocated to either the socioeconomic infrastructure or to market production. According to this treatment, x is the number of adults equivalents employed outside the family. Equivalently, it is the number of "families."

---

[2]The preferences of this model are an example of the $\mathcal{L}^{**}$ lexicographic ordering illustrated in a variety of contexts by Encarnacion. An analysis of the continuity of choice functions based on general $\mathcal{L}^{**}$ preferences is contained in Day (1995). In the present setting $\mathcal{L}^{**}$ preferences mean that adults wish to insure their own survival first (for, of course, the survival of their children depends upon their own); given this they save so as to insure the survival of their children. Given that their children can enjoy a standard of living above the given threshold, parents consider the trade–off between their own and their children's standard of living.

We assume that no production is possible without the appropriate infrastructural resources and that the total capital and total work force cannot be reallocated between production and infrastructure within the current generational period. Substituting (3) and (4) into (2), we get

$$Y = \begin{cases} 0 & ,x \leq \bar{L} \text{ or } Z \leq \bar{K} \\ BF[Z - \bar{K}, x - \bar{L}] & ,x > \bar{L}, Z > \bar{K} \end{cases} \tag{5}$$

Suppose that the infrastructural work force required by the technology is proportional to the total capital stock,

$$\bar{L} = \frac{1}{\nu} Z. \tag{6}$$

This implies that for the capital stock in existence to be productive, the required amount of human capital must be allocated to the infrastructure. Correspondingly, suppose that the capital requirement for the infrastructure depends on the total number of families (= the total work force),

$$\bar{K} = \mu x \tag{7}$$

This implies that for a given population to be productive, the requisite amount of capital must be allocated to the infrastructure. Rewriting (5) gives the production function in the (Z, x) space

$$Y = \begin{cases} 0 & ,(Z, x) \notin C \\ \bar{B}F(Z - \mu x, \nu x - Z) & ,(Z, x) \in C \end{cases} \tag{8}$$

where the *feasibility cone* is

$$C := \{(Z, x) \mid \mu x \leq Z \leq \nu x\} \equiv \{(Z, x) \mid \frac{1}{\nu} Z \leq x \leq \frac{1}{\mu} \leq Z\}.$$

We thus arrive at a Restricted Factor Substitution (RFS) production function called for by Eisner (1993), Kaldor (1959), and Solow (1959), and derived by Day and Zou (1994).

Dividing through both sides of (8) by x and taking advantage of the homogeneity of $F(\cdot)$, we get the production function in per capita terms,

$$Y = \begin{cases} 0 & ,z \in Z^0 := \backslash (\mu, \nu) \\ \bar{B}F(Z - \mu x, \nu x - Z) & ,z \in Z^f := (\mu, \nu) \end{cases} \tag{9}$$

where the production capital/work force ratio is $z = Z/x$ and the output/total work force ratio is $y = Y/x$. Consequently, the infrastructure determines the range of feasible capital/labor ratios, $Z^f := (\mu, \nu)$.

Given the assumptions stated above, routine calculations show that $f(\cdot)$ is twice differentiable, and strictly concave for the feasible capital ratios, that is,

$$f''(z) < 0, \quad z \in Z^f, \tag{10}$$

and that

$$\lim_{z \to \mu} f\prime(z) = \infty, \quad \lim_{z \to \nu} f\prime(z) = -\infty \tag{11}$$

Zou (1991, pp. 58–60) showed that for any given output level, the efficient capital/labor combinations lie in a cone, say C^*, within the feasibility cone C, but the isoquants on which these combinations lie approach the boundaries of C asymptotically and extend outside C^*. In other words, $C \setminus C^*$ contains capital/labor ratio combinations that are feasible but inefficient. The rays defining C^* are determined by constants, say μ^*, ν^* with $\mu < \mu^* < \nu^* < \nu$ which defines the interval of efficient capital/labor ratios

$$\mathcal{Z}^* := [\mu^*, \nu^*],$$

evidently, $f'(z) \geq 0$ for all $z \in \mathcal{Z}^*$.

Using the Cobb–Douglas Production Function

$$Y = BF(K, L) = BK^\beta (L)^{1-\beta} \tag{12}$$

the RFS production function is

$$Y = \begin{cases} 0 & , (Z, x) \notin C \\ B (Z - \mu x)^\beta (\nu x - Z)^{1-\beta} & , (Z, x) \in C \end{cases} \tag{13}$$

where $\bar{B} = B\nu^{\beta-1}$. In terms of the capital/labor ratio,

$$y = \bar{B} f(z) = \begin{cases} 0 & , z \notin \mathcal{Z}^0 \\ \bar{B} (z - \mu)^\beta (\nu - z)^{1-\beta} & , z \in \mathcal{Z}^f. \end{cases} \tag{14}$$

The feasibility cone $\mathcal{Z}^f = (\mu, \nu)$. The rays define the efficiency cone are given by

$$\mu^* = \frac{\mu\nu}{\beta\mu + (1-\beta)\nu} \quad \text{and} \quad \nu^* = \beta\nu + (1-\beta)\mu.$$

Figure 2 illustrates the isoquants of the RFS production function (13).

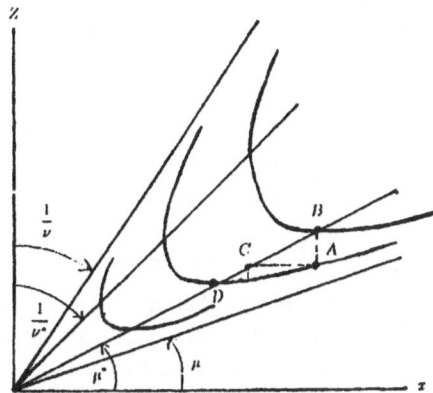

Figure 2: Isoquants of the RFS Production Function

Consider the factor combinations shown. Point A lies outside the efficiency cone. The same amount of output could be obtained by a combination D. Such a point would exhibit excess capacity and unemployment, but it would necessarily involve free disposal of capital and labor. Points C involves an increase in output which could be attained by reallocating some production capital to the infrastructure. This would require fungibility in the short run and, although excess capacity would be eliminated, some labor would have to be unemployed. Point B would involve the same increased output level as points A and C but with full employment. It would require an increase in production capital and the reallocation of part of the labor force to the infrastructure.

All these of those moves from A to B, C or D would require a complete knowledge of the production function and the ability to exploit it within the current time period. They would also require a complete fungibility of human and production capital in the short run, that is, within the current production period.

Interpreted literally, this means that if the political and economic administrators and managers could perceive and understand society's production structure as a whole. More production and a higher standard of living could be achieved for any given combination outside of C^* by rearranging capital and labor so as to obtain a combination within C^*. It seems to us unrealistic to suppose that they always manage to do this. In any case, we consider the situation where they do not.[3]

---

[3]Infrastructure has always been considered by development economists and is now receiving increasing emphasis as reflected, for example, in a recent World Bank (1995) report. That report makes it clear that many countries have in the past failed to invest sufficiently or effectively in adequate infrastructure to support growth. There is a growing body of empirical work aimed at estimating

4 The Consumption/Savings Strategy

The potential sustainable future standard of living anticipated by the present generation must account for the maintenance of the capital stock and an endowment for net additions to the family, so

$$c^1 = y^1 - (n + \delta)z^1 \qquad (15)$$

where y^1 is anticipated sustainable income level for the next generation, z^1, is the planned future capital stock. The capital bequest sufficient to sustain this standard of living is equal to the amount of capital remaining at the end of the period plus sufficient investment to allow for population growth. Therefore,

$$z^1 = \frac{1}{1+n}[y - c + (1 - \delta)z]. \qquad (16)$$

Suppose our forward looking agents perceive current output y and the current rate of return r. Suppose further that they use the latter as a basis for the consumption/bequest trade–off. If r^1 is the rate of return anticipated for the next period, then, in effect,

$$r^1 = r.$$

Our boundedly rational agents then get a rough estimate or "first–order approximation" of y^1 by setting

$$y^1 = y + r(z^1 - z) \qquad (17)$$

Given these assumptions, the consumer's budget set is

$$B := \left\{ (c, c^1) \mid \rho c + c^1 \leq (1 + \rho)[y - (n + \delta)z], \ c \geq 0, c^1 \geq 0 \right\}$$

where, for notational convenience,

$$\rho = \frac{r - (n + \delta)}{1 + n}.$$

Obviously, ρ can be positive, zero, or negative, depending on whether or not r is greater than, equal to, or less than $(n + \delta)$.

The most preferred combination of present consumption and the anticipated future sustainable standard of living, (c, c^1) must satisfy the maximizing relationship,

$$V(y, z, \rho) := \max_{(c,c^1) \in B} u(c, c^1) \qquad (18)$$

---

its contributions. Some references are given in Day and Zou (1994).

Clearly, the optimal current consumption is a function of z, y, and ρ, which we denote by $c(z, y, \rho)$.

It should be emphasized that our boundedly rational agents do not account for the possibility that each succeeding generation will face a similar problem. Each generation will weigh the opportunity cost and satisfaction of current savings given their inherited wealth, their current income and the current rate of return on savings. Each bequest is, therefore, merely a potential consumption level which could be enjoyed forever if each succeeding generation chose to maintain the stock of capital at its inherited level. In effect each generation bestows a flexible asset, the potential income in perpetuity, leaving it up to its heirs to decide what to do with it when they take control.

Given the lexicographic utility function (1), we have

Proposition 1 *Current consumption based on agents' adaptive behavior is given by*

$$c(z, y, \rho) = \begin{cases} y & \text{if } y \leq \bar{c} \text{ or } \rho < 0 \\ \bar{c} & \text{if} \bar{c} < y \leq \bar{c} + (n+\delta)\bar{c} \text{ or } \rho > 0 \\ \min\{y, g(z, y, \rho)\} & \text{if } \bar{c} + (n+\delta) < y \text{ or } \rho > 0 \end{cases}$$

Proof It is obvious that $c(z, y, \rho) = y$ for $y \leq \bar{c}$. Moreover, consumption in the first period should be as big as possible if $\rho \leq 0$, i.e., $c(z, y, \rho) = y$ for $\rho < 0$.

If $\rho > 0$, $y > \bar{c}$, but $c^1 = (1+\rho)[y - (n+\delta)z] - \rho\bar{c} \leq \bar{c}$ or $y - (n+\delta)z \leq \bar{c}$, then $c(z, y, \rho) = \bar{c}$. If $\rho > 0$ and $y - (n+\delta)z > \bar{c}$ and $\lim_{c \to 0} w'(c) = +\infty$, then $w'(c - \bar{c}) - \psi\rho w'(c^1 - \bar{c})$ is a decreasing function of c for $c \in \left[\bar{c}, \frac{(1+\rho)[y-(n+\delta)z]-\bar{c}}{\rho}\right]$. It approaches $+\infty$ as c approaches $\bar{c}$ and approaches $-\infty$ as c approaches $\frac{(1+\rho)[y-(n+\delta)z]-\bar{c}}{\rho}$. Therefore, for each (y, z, ρ) there is a unique solution to the following first order condition

$$w\prime(c - \bar{c}) = \psi\rho w'(c^1 - \bar{c}) \tag{19}$$

which determines a continuous single valued map $c = g(y, z, \rho)$. Since $g(y, z, \rho)$ is not necessary less than y,

$$c(y, z, \rho) = \min\{g(y, z, \rho), y\} \quad \text{if} \quad \rho > 0 \quad \text{and} \quad y - (n+\delta)z > \bar{c}. \blacksquare$$

Our model's agents are assumed to observe their income wealth and the marginal product of capital directly and use this information in their decisions. In theory, however,

$$y = Bf(z), \qquad r = Bf\prime(z) \tag{20}$$

Consequently, (unknown to our agents) the adaptively optimal consumption depends on the current capital stock,

$$c = h(z) := c \left(Bf(z), z, \frac{Bf\prime(z) - (n+\delta)}{1+n} \right) \tag{21}$$

We call this the *implicit consumption strategy*.

Proposition 2 *The Implicit Consumption Strategy.* Assume there exits a value of z such that $Bf(z) - (n+\delta)z > \bar{c}$. Then

(i) There exist capital/work force ratios, z', z'', z''' satisfying $\mu < z' < z'' < z''' < \nu$ and there exists a positive, differentiable, function, $g : \mathbb{R}^+ \to \mathbb{R}^+$ such that

$$h(z) = \begin{cases} Bf(z) & , \; z \in Z' := [\mu, z') \\ \bar{c} & , \; z \in Z'' := [z', z'') \\ \min\{y, g(z)\} & , \; z \in Z^s := [z'', z''') \\ Bf(z) & , \; z \in Z''' := [z''', \nu] \end{cases}$$

(ii) there exists a total capital total work force ratio, $\tilde{z} \in c\ell Z^s$ such that

$$\tilde{\rho} = \frac{\tilde{r} - (n+\delta)}{1+n} = \frac{1}{\psi} \quad \text{(equivalently} \quad \tilde{r} = \frac{(1+n)}{\psi} + (n+\delta) \text{)}.$$

Proof From Proposition 1 we can give an explicit form for $h(z)$. There must exist a unique capital/labor ratio z''' such that $\mu < z''' < \nu$ and $Bf'(z''') = (n+\delta)$. If $Bf(z''') - (n+\delta)z''' > \bar{c}$, clearly there exist z', z'' with $\mu < z' < z'' < z''' < \nu$ such that $Bf(z') = \bar{c}$ and $Bf(z'') - (n+\delta)z'' = \bar{c}$. See figure 3.

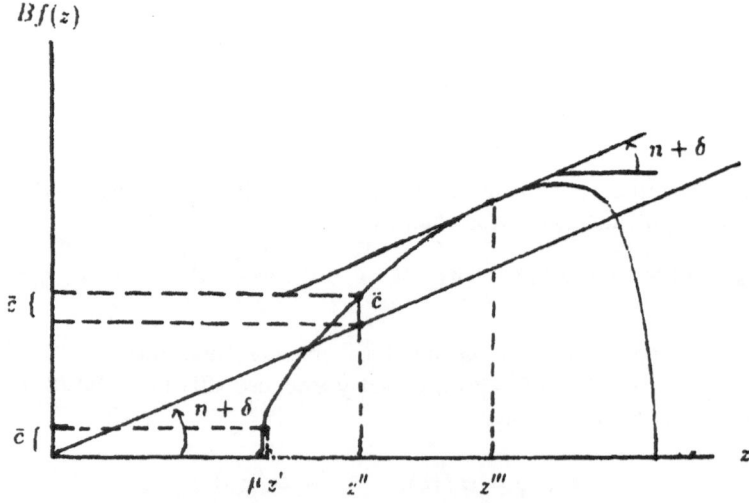

Figure 3: Critical Capital/Labor Ratios

It is easy to see that

$$
\begin{array}{ll}
\rho < 0 & \text{for all } z > z''' \\
y < \bar{c} & \text{for all } \mu < z < z' \\
y > \bar{c}, \ y - (n+\delta)z < \bar{c} \text{ and } \rho > 0 & \text{for all } z' < z < z'' \\
y - (n+\delta)z > \bar{c} \quad \text{and} \quad \rho > 0 & \text{for all } z'' < z < z'''
\end{array}
$$

Given these facts, the results follow. ■

5   Capital Accumulation

Given a population growth rate, n, and a depreciation rate, δ, and given the adaptive optimizing consumption strategy, the behavior of the capital/labor ratio is governed by

$$
z_{t+1} = \theta(z_t) = \frac{1}{1+n}[(1-\delta)z_t + Bf(z_t) - h(z_t)].
$$

Given the piecewise nature of $h(z)$, the map $\theta(\cdot)$ has a piecewise, multiple phase structure,

$$
z_{t+1} = \theta(z_t) = \begin{cases}
\theta_d(z_t) := \frac{(1-\delta)}{1+n} z_t & , z_t \in Z' \\
\theta_e(z_t) := \frac{1}{1+n}[(1-\delta)z_t + Bf(z_t) - \bar{c}] & , z_t \in Z'' \\
\theta_s(z_t) := \frac{1}{1+n}[(1-\delta)z_t + Bf(z_t) - h(z_t)] & , z_t \in Z^s \\
\theta_d(z_t) := \frac{(1-\delta)}{1+n} z_t & , z_t \in Z'''
\end{cases}
\tag{22}
$$

which is illustrated in figure 4.

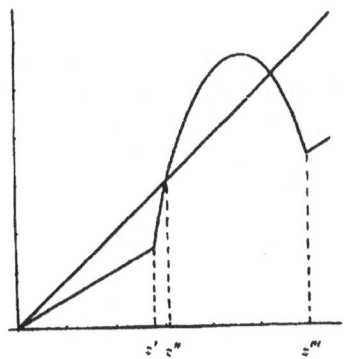

Figure 4: The Dynamic Structure

Recalling our notational convention, the capital stock planned at time t for time $t+1$, $z_t^1 = z_{t+1}$. But $y_{t+1} \neq y_t^1$ except at a steady state. Consequently, $c_t^1 \neq c_{t+1}$ except at a steady state.

Proposition 3 *Existence of Stationary States. There exists a unique positive interval of (B_0, B^0), such that for all $B \in (B_0, B^0)$ there are two distinct positive, steady states, $\tilde{z}^l, \tilde{z}^\mu$ with $\tilde{z}^l < \tilde{z}^\mu$.*

Proof Let z^0 be defined by $Bf'(z^0) - (n+\delta) = 0$, so that $\rho(z^0) = 0$. Let $\bar{B}$ be defined by $\bar{B}f(z^0) - (n+\delta)z = \bar{c}$, so that $z'' = z^0$. As B increases, z'' decreases and $f'(z'')$ increases. $B \to \infty$ implies $\rho(z'') \to \infty$. According to the Intermediate Value Theorem, there exists a B_0 such that

$$\frac{B_0 f'(z'') - (n+\delta)}{1+n} = \frac{1}{\psi}. \tag{23}$$

If $B > B_0$, then

$$\frac{Bf'(z'') - (n+\delta)}{1+n} > 1/\psi \tag{24}$$

so that $\tilde{z} > z''$.

Consider $\tilde{z} < z < z^0$, we know that $0 < \frac{Bf'(z)-(n+\delta)}{1+n} < 1/\psi$. As long as B is sufficiently large, (26) is true for all $z \in (\tilde{z}, z^0)$. Thus, $\rho(z''') = 1/\psi$ and $\tilde{z} = z'''$. Let B^0 be the smallest value of B such that (26) is true for all $z \in (\tilde{z}, z^0)$. Thus, if $B < B^0$, then $\tilde{z} < z'''$. Therefore, for any $B \in (B_0, B^0)$, $\tilde{z} \in Z^s$. But we know that z'' is a steady state where $c = \bar{c}$ and $c^1 = \bar{c}$. Thus, there exist two distinct steady states, $\tilde{z}^l$ and $\tilde{z}^\mu$, where $\tilde{z}^l = z''$ and $\tilde{z}^\mu \in Z^s$. When $B = B_0$, there exists unique $\tilde{z} = z''$. When $B < B_0$, no steady state exists. When $B \geq B^0$, $\tilde{z} = z'''$ which is no longer a steady state since $c(z''') = Bf(z''')$. ■

Proposition 4 *Boundedness of Trajectories in the Capital/Work Force Ratio. The smaller steady state is unstable and all trajectories $\tau(z) < \tilde{z}^l$ converge to zero and all trajectories $\tau(z)$ such that $\tilde{z}^\mu > z > \tilde{z}^l$ grow for at least one period. Given $n, \delta > 0$, there exist constants t_s, z^m, z^M such that for all $z > \tilde{z}^l$ and for all $t > t_s$,*

$$\theta^t(z) \in [z^m, z^M]$$

where $z^m = \theta(z''')$ and $z^M = \arg\max_{z \in Z^s} \theta(z)$.

Proof If the initial capital labor ratio is smaller than $\tilde{z}^l (= z'')$, then $c = \bar{c}$ and $c^1 < \bar{c}$. If $z \in Z'$, then $z_{t+1} = \left[\frac{1-\delta}{1+n}\right]^t z_1 \to 0$ as $t \to \infty$.

If $z' < z_1 < z''$, then

$$c^1\left(\bar{c},\, Bf(z_1),\, z_1,\, \frac{Bf'(z_1) - (n+\delta)}{1+n}\right) < \bar{c},$$

which implies that $s(z_1) < (n+\delta)z_1$. Thus,

$$z_2 = \frac{s(z_1) + (1-\delta)z_1}{1+n} < z_1.$$

Similarly, we have $z_3 < z_2$, $z_4 < z_3, \ldots, z_{t+1} < z_t$. As long as $z' < z < z''$, we have $s(z) = Bf(z) - \bar{c}$. Thus,

$$z_1 - z_2 = \frac{(n+\delta)z_1 - Bf(z_1) + \bar{c}}{1+n},$$

$$z_2 - z_3 = \frac{(n+\delta)z_2 - Bf(z_2) + \bar{c}}{1+n}.$$

Remember that $\rho(z_1) > 0$, so that $Bf'(z_1) > (n+\delta)$, we have

$$\frac{Bf(z_1) - Bf(z_2)}{z_1 - z_2} > Bf'(z_1) > n + \delta.$$

Rearrange the terms in the above inequality, we have

$$(n+\delta)z_1 - Bf(z_1) < (n+\delta)z_2 - Bf(z_2).$$

Thus, we have $z_1 - z_2 < z_2 - z_3$. Similarly we can prove $z_2 - z_3 < z_3 - z_4 < \ldots$. But

$$z_1 - z_{t+1} = (z_1 - z_2) + (z_2 - z_3) + \cdots + (z_t - z_{t+1}) > t(z_1 - z_2).$$

Therefore, for an initial z_1 such that $z' < z_1 < z''$, there exists a sufficiently large t such that $z_{t+1} \in Z'$. This means that for any initial $z_1 < z''$, z will eventually be reduced to such a level that savings are no longer positive. Therefore, $z \to 0$ as $t \to \infty$.

If $\tilde{z}^\mu > z_t > \tilde{z}^l$, we know that $s(z) > (n+\delta)z$. This implies that $z_{t+1} > z_t$. ∎

Proposition 5 *Generic Asymptotic Stability and Generic Instability. With certain weak restrictions on technology, for all positive n, δ and $B^0 > B > B_0$, there exist open sets (consisting of a union of open intervals) $\mathcal{F}^{ms}, \mathcal{F}^{cs}, \mathcal{F}^{\mu s}$ such that (i) for all $\psi \in \mathcal{F}^{ms}$ the steady state is asymptotically stable and is approached monotonically; (ii) for all $\psi \in \mathcal{F}^{cs}$ capital/labor ratio trajectories eventually exhibit dampening fluctuations converging to the steady state; for all $\psi \in \mathcal{F}^{\mu s}$ almost all trajectories* *(except those that hit $\tilde{z}$ after a finite number of periods) eventually exhibit persistent fluctuations around the steady state.*

Proof Since $g(z)$ satisfies

$$w'(g(z) - \bar{c}) = \psi \rho w'(c^1 - \bar{c})$$

where $c^1 = (1 + \rho)\,[y - (n + \delta)z] - \rho c$. This implies

$$[w''(g(z) - \bar{c} \quad + \quad \psi \rho^2 w''(c^1 - \bar{c})]\frac{dg(z)}{dz}$$
$$= \quad \psi \rho' w'(c^1 - \bar{c}) + \psi \rho w''(c^1 - \bar{c})$$
$$\times \{\rho'[y - (n + \delta)z - c] + \rho(\rho + 1)(1 + n)\}.$$

At the steady state $g(z) = c^1 \implies g(\tilde{z}) = \tilde{c} = \tilde{y} - (n + \delta)\tilde{z}$ and $\tilde{\rho} = \frac{1}{\psi}$. Therefore,

$$\frac{dg(\tilde{z})}{dz} = \frac{1 + n}{\psi} + \frac{\psi}{1 + \frac{1}{\psi}}\tilde{\rho}'\frac{w'(\tilde{c} - \bar{c})}{w''(\tilde{c} - \bar{c})}$$
$$\implies \theta'(\tilde{x}) = 1 - \frac{\psi}{(1 + n)^2(1 + \frac{1}{\psi})}\tilde{r}'\frac{w'(\tilde{c} - \bar{c})}{w''(\tilde{c} - \bar{c})}.$$

Since $\tilde{r} = (n + \delta) + \frac{1}{\psi} \implies \lim_{\psi \to \infty} \tilde{z}(\psi) = z^* \implies \lim_{\psi \to \infty} \theta'(\tilde{z}) = -\infty$, it is easy to see that $\theta'(\tilde{z})$ is a continuous function of ψ and $\theta'(\tilde{z}) < 1$ for all $\psi > 0$. Let

$$\mathcal{F}^{ms} = \{\psi \mid 0 < \theta'[\tilde{z}(\psi)] < 1\}$$
$$\mathcal{F}^{cs} = \{\psi \mid -1 < \theta'[\tilde{z}(\psi)] < 0\}$$
$$\mathcal{F}^{\mu s} = \{\psi \mid \theta'[\tilde{z}(\psi)] < -1\},$$

then the proposition holds. ∎

The fluctuations may be cyclic or chaotic and may be strongly ergodic if the Miseurwcz condition pertains.

6 Example

Assume the utility function is defined by (2) and

$$w(c) = A(c - \bar{c})^\alpha, \quad 0 < \alpha < 1 \tag{25}$$

The first order condition when both present and potential future consumption are both positive is

$$\frac{1}{\psi} \cdot \left(\frac{c^1 - \bar{c}}{c - \bar{c}}\right)^{1-\alpha} = \rho \tag{26}$$

Substituting for c^1, using (18) and doing some rearranging, it follows that

$$c = g(z) = \frac{(\rho\psi)^{\frac{1}{\alpha-1}}\{(1+\rho)[y - (n+\delta)z] - \bar{c}\}}{1 + \rho(\rho\psi)^{\frac{1}{\alpha-1}}} \tag{27}$$

This function, we recall, is defined on $z \in Z^s$. This gives the phase structure when savings are positive:

$$\theta_s(z) = \frac{1}{1+n}\left\{(1-\delta)z + y - \frac{(\rho\psi)^{\frac{1}{\alpha-1}}\{(1+\rho)[y - (n+\delta)z] - \bar{c}\}}{1 + \rho(\rho\psi)^{\frac{1}{\alpha-1}}}\right\} z \in Z^s \tag{28}$$

In figure 5 capital/labor and income trajectories are shown.

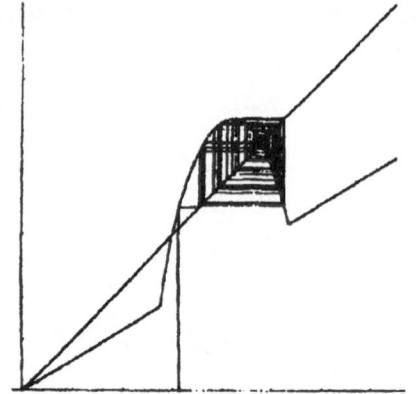

(a) The Capital/Labor Ratio in Phase Space

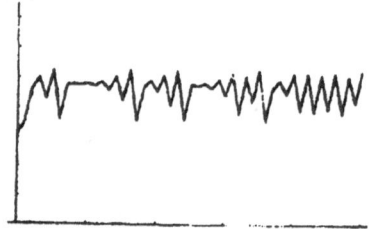

(b) The Capital/Labor Ratio Trajectory

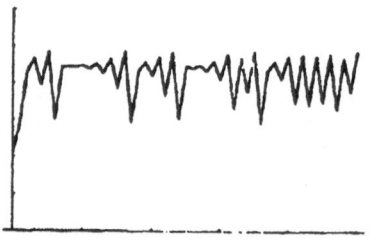

(c) The Per Family Income Trajectory

Figure 5: Sample Trajectories

References

[1] Boserup, Ester, 1981, *Population and Technological Change, A Study of Long Term Trends*, The Chicago University Press, Chicago.

[2] Day, Richard H. and Jean–Luc Walter, 1989, "Economic Growth in the Very Long Run: On the Multiple–Phase Interaction of Population, Technology, and Social Infrastructure," Chapter 11 in W. Barnett, J. Geweke, K. Shell (eds.), *Economic Complexity: Chaos, Sunspots, Bubbles and Nonlinearity*, Cambridge University Press, Cambridge.

[3] Day, Richard H. and Gang Zou, 1994, "Infrastructure, Restricted Factor Substitution and Economic Growth," *Journal of Economic Behavior and Organization*, 23, 149–166.

[4] Day, Richard H., 1996, "Satisficing Multiple Preferences In and Out of Equilibrium," in R. Fabella and E. de Dios (eds.), *Choice, Growth and Development: Emerging and Enduring Issues*, University of the Philippines Press, The Philippines.

[5] Eisner, Robert, 1993, A Presentation paper at the Annual Meeting of the American Economic Association, Los Angeles.

[6] Grandmont, Jean–Michel and Guy Laroque, 1973, "Money in a Pure Consumption Loan Model," *Journal of Economic Theory*, 6, 382–395.

[7] Grandmont, Jean–Michel and Guy Laroque, 1986, "Stability of Cycles and Expectations," *Journal of Economic Theory*, 40, 138–151.

[8] Grandmont, Jean–Michel and Guy Laroque, 1990, "Economic Dynamics with Learning," in *Equilibrium Theory and Applications*, Cambridge University Press, Cambridge.

[9] Kaldor, Nicholas, 1959, "Economic Growth and the Problem of Inflation," *Economica*, 26, 287–298.

[10] King, Robert G., and Sergio Bebelo, 1990, "Government Spending in a Simple Model of Endogenous Growth," *Journal of Political Economy*, 98, 126–150.

[11] North, Douglass C., 1981, *Structure and Change in Economic History*, W.W. Norton and Company, New York.

[12] Solow, Robert M., 1956, "A Contribution to the Theory of Economic Growth," *Quarterly Journal of Economics*, 70, 65–94.

[13] Solow, Robert M., 1959, "Is Factor Substitution a Crime? If So, How Bad? Reply to Professor Eisner," *Economic Journal*, 69, 597–599.

[14] World Bank, 1994, *Infrastructure for Development: World Development Report*, Oxford University Press, New York.

[15] Zou, Gang, 1991, *Growth with Development*, Ph.D. Dissertation, University of Southern California, Los Angeles.

Neo-Classical Growth and Complex Dynamics: A Note on Day's (1982) Model

Gilles DUFRÉNOT

1 Introduction

Considerable works have been done on chaotic dynamics in the field of economic growth and dynamic macroeconomics (see Day and Gang [1], Day [2], Nishimura and Yano [3], Grandmont et al. [4]. The study of chaotic dynamics in economic growth has its root in a paper dating back from 1982 by Richard Day. Our purpose is to consider new aspects of this original contribution. Day's [5] established the existence of a chaotic growth due to the presence of a "pollution effect" in the capital stock accumulation process. Two aspects of his paper are under discussion here.

Firstly, the author considered a "pollution effect" to illustrate the possible existence of a negative externality in the aggregate capital stock accumulation process. This lead him to assuming a production function for the representative firm of the economy, having a humped concave shape. However, as argued by Lorenz [6, pp. 123-125], Day's contribution may be viewed as a prototype of a model that can exhibit chaotic dynamics with an *ad hoc* modification of the standard textbook approach to neoclassical growth theory. In addition to the *ad hoc* nature of the "pollution effect", other authors have commented on the lack of any microeconomic foundations in Day's model. Because of this, one may cast some doubts on all the models that have been developed in the literature and which use the same methodology. These criticisms which raise methodological problems, have received no answers in the literature. It is the aim of this note to provide one plausible explanation that can be used to justify the presence of the "polluting effect".

It will be shown that the latter can be interpreted within the framework of a disaggregate economy, as a consequence of the heterogeneity amongst entrepreneurs, notably in their expectations behavior.. In this context, the formal content given to the "pollution effect" in Day's model is an aggregate indicator of firms interactions, when entrepreneurs try to coordinate their decisions to accumulate capital and have no knowledge of the level of global demand. This formulation of Day's model suggests that volatile expectations may be a dominant cause of growth irregularity in a one-sector economy, when commodity markets are imperfect.

This reinterpretation thus leads to consider the question of the microeconomic foundations of the "pollution effect" that is responsible for the presence of a chaotic dynamics. It will become apparent in this study that volatile expectations in a decentralized economy may have persistent effects, in aggregate, causing the instability of economic growth. In other words, there exists a close connection between the non-linear specification representing the aggregate capital accumulation and the process that leads firms to adjust their capital stock according to the decisions taken by their competitors. The idea of locating the nonlinearity that drives Day's dynamic results in expectations interactions at the firm level was the subject of an earlier paper by Albin [7]. Even thought our discussion may seem close to that of this author, the approach that we follow is rather different. Albin stood exclusively at the disaggregated level and did not compare the properties of the implications of his model with those at the macroeconomic level. He rather used a microeconomic cellular automaton arguing that the process of aggregation eliminates certain dynamic evolutions observed at the individual level. Our contribution consists in showing that the complex expectations interactions can be aggregated to a single function of the capital-labor ratio just similar to the one analyzed in Day's paper.

The second element which is under discussion deals with some mathematical points concerning the concept of complex dynamics used in Day's model. It is now agreed that one essential feature of a complex or chaotic dynamics is the sensitivity dependence on initial values with positive Lebesgue measure (this is ergodic chaos). Besides, following Day [8] and Melese and Transue [9], two other notions have emerged in the literature, namely "thick chaos" and "thin chaos". The first concept corresponds to "true" chaos and is associated with behavior switches from stable periodic orbits to non periodic time-paths. The notion of thin chaos leads to consider that chaos is not truly observed despite the fact that there exists high-order cycles that are indistinguishable from aperiodic fluctuations.

In this paper, we discuss another aspect of complex dynamics. By using the bifurcation analysis, it is shown that the solutions of non-linear difference equations that produces irregular fluctuations are highly sensitive to a variation of the parameters. These solutions change erratically for certain ranges of parameter values. What this implies is that even when the conditions for a stable growth are present, transient chaos can be observed and then it may take the economy a (very) long time to converge towards a sequence of capital accumulation which is monotone over time.

The paper is organized as follows. In section 2, the model is presented and we provide some theoretical justifications of its equations. Our modified formulation of Day's model mainly concerns the assumptions about expectations. In section 3, we discuss definitions and procedures for the study of the model stability. One essential feature of this section is that the complexity of the dynamics comes from the fact that the solutions are highly sensitive to a variation of parameters.

2   The modified model

We will set up the model and then explain its equations.

Notations:

s_t: volume of saving at time t

I_{it}: investment of firm i at time t

K_{it} : capital stock of firm i at time t

L_{it}: quantity of labor used by firm i at time t

Y_{it} : production of firm i at time t

s : saving rate

A_{jt} : set of firms j whose decisions have an influence upon firm i's choices

$g_{it}(A_{jt})$: reaction function of the ith firm faced with its j competitors

K_t : capital stock of the economy at time t

L_t: quantity of labor used at time t

Y_t: quantity of good produced by the economy at time t

k_t : capital-labor ratio

The model is summarized by the following equations:

$$Y_{it} = F(K_{it}, L_{it}), \forall i = 1, ... N \tag{1}$$

$$I_t = K_{t+1} \tag{2}$$

$$g_{it}(A_{jt}) = \alpha_i sgn(\Delta K_{it}), \alpha_i > 0 \tag{3}$$

$$K_{t+1} = sF(K_t, L_t)\left[1 + n_t\right], \ n_t = (1/N) \sum_{i=1}^{N} g_{it}(A_{jt}) \tag{4}$$

$$f(k_t) = Ak_t^\beta, \ n_t = h(k_t) = -k_t, \ 0 \le k_t \le 1 \tag{5}$$

$$k_{t+1} = Bk_t^\beta(1 - k_t), \ k_0 > 0, \ B = sA \tag{6}$$

Theoretical justifications of the equations

(1) This simply means that the economy is composed of N firms. They produce Y_{it} units of a good with K_{it} units of capital and L_{it} units of labor.. F is assumed to be degree one homogeneous. This assumption means that the number of firms is of no consequence on the results. All the firms produce the same good and the time delay for production is one period.

(2) For simplicity, we assume that there is no physical depreciation of capital. K_{t+1} is therefore net investment.

(3) Information is imperfect so that no firm can forecast the level of future demand. We assume that when confronted to an uncertainty on the future level of demand, entrepreneurs form their expectations in the following way. Each individual who observes that many of its competitors are increasing (or decreasing) their capital stock is led to believe that they are forecasting a high level of activity for the next period. The only way for him to benefit from this increase (or to avoid over- production) is to produce more (or less) and consequently to increase (or decrease) the capital stock that he invests in the process of production. Each firm i is supposed to react to the decisions of j competitors ($j < N$). The sign function sgn(x) has three values depending upon the sign of x. These values are respectively 1, -1, and 0, depending upon whether x is positive, negative or equals 0. This function is used to formalize the idea that expectations are guided by the collective psychology of the firms. If an entrepreneur observes that other competitors have decided to produce more (and thus are investing more capital today to increase the level of production tomorrow), a positive externality can result from the fact the entrepreneur believes that he will benefit from a stronger demand, by which higher revenue he is guaranteed and more profits. So, increasing concentrations of capital can result from the fact that an entrepreneur is led to invest an amount of capital greater than the amount he would have invested had he no paid attention to the decisions of other entrepreneurs.

(4) The function n_t is a formal description of the way firms coordinate their expectations in aggregate. The usual good market equilibrium condition requires that the capital stock at time t (or net investment) be equal to the aggregate saving, that is equation (4) with $n_t = 0$. The modification we introduce here is that the capital stock is increased (if $n_t > 0$), or decreased (if $n_t < 0$) by an amount that depends on the way firms coordinate their decisions to accumulate capital. This reflects the consequence in aggregate of the behavioral interactions which are due to the crossed expectations of agents. The simplest way to obtain equation (4) is to proceed as follows.

The capital stock accumulation at a given period for an individual firm has two components: firstly, the quantity of good that has not been supply on the commodity market (which depends on the saving rate of consumers); secondly a component which can be interpreted as an externality because firms are supposed to adjust their production according to the decisions taken by their competitors. We therefore write the following relations:

$$K_{it+1} = sY_{it}\left[1 + g_{it}(A_{jt})\right] \tag{7}$$

or

$$K_{it+1} = sF(K_{it}, L_{it})\left[1 + g_{it}(A_{jt})\right] \tag{8}$$

Assuming that all firms employ the same quantity of capital and labor, we write

$$F(k_{it}, L_{it}) = (1/N)F(K_t, L_t) \text{ where } K_t = NK_{it}, \ L = NL_{it} \tag{9}$$

Then, we sum up the firms on each side of (8) and thus get:

$$\sum_{i=1}^{N} K_{it+1} = NK_{it+1} = K_{t+1} = \left[(s/N)F(K_t, L_t)\right]\left[\sum_{i=1}^{N} 1 + \sum_{i=1}^{N} g_{it}(A_{jt})\right] \tag{10}$$

or

$$K_{t+1} = sF(K_t, L_t)[1 + (1/N)\sum_{i=1}^{N} g_{it}(A_{jt})] \tag{11}$$

By examining equation (11), we see that we have to consider the problem of aggregation of the functions $g_{it}(A_{jt})$. To start with a simple example consider an economy composed of four firms. Suppose that in observing each other, three of them decide to reduce their capital accumulation because they forecast a decrease in the level of activity for the next period. Assume further that the fourth firm decides not to modify its rhythm of accumulation. These hypotheses leads to the following relations:

$$g_{1t}(A_{jt}) = \alpha_1 sng(\Delta K_{1t}) = -\alpha_1, \ j = 2, 3, 4 \tag{12}$$

$$g_{2t}(A_{jt}) = \alpha_2 sng(\Delta K_{2t}) = -\alpha_2, \ j = 1, 3, 4 \tag{13}$$

$$g_{3t}(A_{jt}) = \alpha_3 sng(\Delta K_{3t}) = -\alpha_3, \ j = 1, 2, 4 \tag{14}$$

$$g_{4t}(A_{jt}) = \alpha_4 sng(\Delta K_{4t}) = 0, \ j = 1, 2, 3 \tag{15}$$

From the relations above, we have

$$(1/4) \sum_{i=1}^{4} g_{it}(A_{jt}) = -(1/4) \sum_{i=1}^{3} \alpha_i < 0 \qquad (16)$$

In the general case, if we note the average aggregate function by n_t, this term will be negative if the entrepreneurs, taken as a whole, are pessimistic about the level of activity for the forthcoming period (this is the case in our example). Conversely, a positive value will indicate moods of optimism.

(5) For illustration we choose a Cobb-Douglas production function as is frequently used in neoclassical one-sector growth models. Since the functional $F(K_{it}, L_{it})$ is independent of i, we assume that $f(k_t)$ is the reduced production function of the economy. The per-capita production is a continuous function of the capital-labor ratio and it has continuous derivatives of all required orders for $k_t > 0$. In particular, the marginal productivity of capital is a decreasing function of k_t, while the marginal productivity of labor is increasing with k_t. This implies that $f'(k_t) > 0$ and $f''(K_t) < 0$.

Furthermore, it is noteworthy that n_t is dependent on both the value of k_t and its sign. As one can easily understand this variable does fluctuate following the variation of aggregate expected demand. Indeed, since we assume that return to scale are constant, the supply of good and the demand for factors of the firms are a priori undetermined (they vary between 0 and infinity). In other words, there exists an infinity of capital-labor ratio corresponding to the optimal profit. Production is therefore determined by the level of aggregate demand that entrepreneurs forecast. Those try to find the proof of their expectations in the decisions of all others and mechanisms of self fulfillment of representations take place. If a majority of firms are pessimistic, then they will anticipate a decrease in demand in comparison to its present level. Their expectations will realize itself on its own: to avoid loss from the predicted decrease, each entrepreneur will reduce his investment and the number of workers; the wages they do not pay will effectively decrease the aggregate demand. If such behaviors are observed, then the sign of will be negative. In order to facilitate the comparison with Day's model, we assume for illustration that $n_t = h(k_t) = -k_t$. This means that during the periods of increasing per-capita output the economy resources tend to go beyond the level of global demand, thereby inducing firms to reduce capital accumulation.

(6) This follows by combining equations (4) and (5) and by dividing each side of equation (4) by . We further assume that the growth rate of the population is 0. The results are unaltered when a positive growth rate is assumed.

Comparison with Day's (1982) formulation

Equation (6) is a particular case of Day's formulation who supposed that the dynamics of the capital-labor ratio was described by the following equation:

$$k_{t+1} = sf(k_t) = s[Ak_t^\beta(m - k_t)^\gamma] = Bk_t^\beta(m - k_t)^\gamma \qquad (17)$$

The term $(m - k_t)^\gamma$ reflects the influence of the "pollution effect" on per-capital output. Increasing concentrations of capital lead to a saturation level m. This means that some potential resources of the economy have to be sacrificed if one wonder to avoid such a saturation.

However, when dealing with equation (17), it is not enough to assume that a "pollution effect" can occur, but one also need to understand the economic factors that explain such an effect. Day's approach may appear ad hoc because the author does not answer the second part of the question. The developments above show how the nonlinearities induced by "polluting effects" can be achieved by investigating the role played by crossed expectations. Expectations are crucial in determining the allocation of production and capital over time: periods of increasing accumulation may cause firms to decrease their investment in subsequent periods, if they believe that the level of future demand will be too low in comparison to the level of production that would be induced by capital increases. Furthermore, this decision is not taken individually. Each entrepreneur's problem is to know and decide whether or not to reduce its investment, in the light of the choice made by the majority of agents, a behavior which is given a formal content through the use of the function $g_{it}(A_{jt})$ and its aggregate counterpart.

One thus observes that even though equations (6) and (17) have identical mathematical formulation (when $m = \gamma = 1$), their economic contents are different. In the formulation above, the mimetic effect due to crossed expectations of entrepreneurs explain why the capital accumulation is described by a logistic equation. Moreover, the production function has the usual Cobb-Douglas form. On the opposite, in Day's approach the evolution of the capital-labor ratio has its roots in the hypotheses formulated on the production technology. However, many authors have argued that in this second case the "pollution effect" lacks some explanations on the sources of this effect. We have tried here to answer this question using a modelling strategy that renders Day's approach more keynesian than was the author's original contribution. As is well known the problem of expectation coordination has been brought to the fore by Keynes to explain the functioning of stock markets. We have simply tried here to extend his arguments to commodity markets. The presence of a chaotic accumulation path can be interpreted from equation (6) as a consequence of imperfections that are due to the organizational structure of the commodity markets. And in this context, expectations of individual firms are guided by their collective psychology, which can be the very motor of instability of fluctuations.

We now discuss another aspect in Day's approach, that also concerns all those models which produce chaotic dynamics from logistic family equations. The reason for such a discussion is the following. There are plenty of accounts in the literature on the properties of sensitivity to initial conditions that characterize complex dynamics. Also, the notion of ergodic chaos has been emphasized. However, little have been said on the highly sensitivity to parameters that characterize chaotic models. This point is important since it helps understanding why *transient* chaos occurs. Transient chaos is produced whenever a stationary (or a low-order periodic cycle) exists as a long term stable equilibrium, but the capital stock behaves like a chaotic series before its trajectory eventually settles down on the trajectory that leads to the stable stationary state (or to the stable limit-cycle). This is the case when the stable solutions coexist with a high number of high-order limit-cycles that are unstable. In this context, the problem of sensitivity to initial conditions may be reformulated as follows: one wonders about the best way for the economy to achieve its stable attractor while not spending most of its time too far from it. Of course, this depends of the initial point. The closer the initial capital-labor ratio is from the stationary state, the higher is the likelihood to achieve the stationary state quickly and the lesser to observe transient erratic fluctuations. These point are now examined in the following section.

3   Sensitivity to parameters and transient chaos in logistic family equations

To investigate the property of sensitivity to parameters in logistic family equations, we present a simple analytical method that enables to show that the subsequent bifurcation points can be ordered in sets which have a "boxes within boxes" structure. The procedure involves four stages.

(1) First, we need a simpler formulation of equation (17).

(2) Secondly, we define the limit bounds of the interval within which the parameters vary.

(3) Thirdly, we show that the existence of chaotic fluctuations is not incompatible with the existence of a period-halving bifurcation.

(4) Fourthly, we consider the problem of highly sensitivity to parameters, when chaotic dynamics coexists with odd periods cycles.

3.1   A REFORMULATION OF EQUATION (17)

We focus here on the case $m = \beta = \gamma = 1$. This case is simpler to analyze than the more complex formulation $m \neq \beta \neq \gamma \neq 1$. This general case is examined in a forthcoming paper.

Proposition 1 *Using a linear change of variable, the dynamics of the capital-labor ratio can be studied by considering the following equation:*

$$x_{t+1} = P(x_t, \mu) = x_t^2 + \mu, \quad \mu = -B^2/4 + B/2 \qquad (18)$$

Proof. Consider the following change of variable:

$$x_t = \phi k_t + \theta \qquad (19)$$

where θ and ϕ are two real parameters. This implies:

$$k_t = \frac{x_t - \theta}{\phi} = (1/\phi)x_t - (\theta/\phi) \qquad (20)$$

and:

$$k_{t+1} = (1/\phi)x_{t+1} - (\theta/\phi) \qquad (21)$$

Equation (17) is thus written:

$$(1/\phi)x_{t+1} - (\theta/\phi) = B\left[(1/\phi)x_t - \theta/\phi\right] - B\left[(1/\phi^2)x_t^2 - 2(\theta/\phi^2)x_t + (\theta^2/\phi^2)\right] \qquad (22)$$

Under this restriction, equation 17 leads:

$$x_{t+1} = Bx_t - B\theta - (B/\phi)x_t^2 + 2(\theta B/\phi)x_t - (B\theta^2/\phi) + \theta \qquad (23)$$

Let us assume that $\phi = -B, \theta = B/2$. Equation (23) becomes:

$$x_{t+1} = x_t^2 + \mu, \text{ where } \mu = -B^2/4 + (B/2) \qquad (24)$$

∎

Remark 1 *Under the assumptions made above on the parameters ϕ and θ the equation (24) is much more simpler to study than equation (17). μ can thereby be considered as the bifurcation parameter that we vary and from which the bifurcation points of the saving rate are deduced.*

Remark 2 *The values of the saving rate are expressed as follows. From equation (17) we know that $sA = B$. Since $\mu = -B^2/4 + (B/2)$, we have $B^2 - 2B + 4\mu = 0$, which implies $B_{1,2} = 1 \pm (1 - 4\mu)^{1/2}$ with $\mu < 1/4$. We deduce that $s_{1,2} = 1/A \pm (1 - 4\mu)^{1/2}/A$. Given the restrictions above, we only consider the greatest solution. We shall assume henceforth that $A = 4$.*

3.2 DETERMINATION OF THE LIMIT BOUNDS OF THE BIFURCATION PARAMETER

Proposition 2 *The smallest value of the saving rate s is $s_{\min} = (1/4)$.*
Proof. *Let us consider the solution $B_1 = 1 + (1 - 4\mu)^{1/2}$. Since $\mu < 1/4$, $B_1 > 1$. This inequality implies:*

$$sA = 1 + (1 - 4\mu)^{1/2} > 1 \tag{25}$$

and $s > (1/A) = s$. If $A = 4$, then the smallest value of the saving rate is $s = 1/4$. ∎

To find the highest value of the saving rate, we first need the following lemma:

Lemma 3 *The successive iterates of the map $P(x_t, \mu)$ are written in the form of two polynomials in x_t and μ.*

Proof. We first calculate x_{t+2} as $P^2(x_t, \mu) = P[P(x_t, \mu)] = (x_t^2 + \mu)^2 + \mu$ and develop this expression:

$$P^2(x_t, \mu) = (x_t^4 + 2\mu x_t^2) + (\mu^2 + \mu = \Phi_2(x_t) + G_2(\mu) \tag{26}$$

$\Phi_2(x_t)$ is a polynomial in x of degree 4 and $G_2(\mu)$ is a polynomial in μ of degree 2.

Let us see how x_{t+3} is written:

$$x_{t+3} = P^3(x_t, \mu) = P[P^2(x_t, \mu)] = [\Phi_2(x_t) + G_2(\mu)]^2 + \mu \tag{27}$$

$\Phi_2(x_t)$ is a 4th degree polynomial and thus the highest degree of $P^3(x_t, \mu)$ is of the same degree as the term $[\Phi_2(x_t)]^2$, that is a polynomial of degree 8. On the other hand, the terms in $P^3(x_t, \mu)$ which are dependent on μ alone are given by the 4th degree polynomial:

$$G_3(\mu) = [G_2(\mu)]^2 + \mu \tag{28}$$

Thus $P^3(x_t, \mu)$ is written as the sum of two polynomials, one in x_t, of degree 8 and the other in μ of degree 4.

Suppose that the previous relations are satisfied in the order k-1:

$$x_{t+(k-1)} = P^{k-1}(x_t, \mu) = x_t^{2^{k-1}} + ... + G_{k-1}(\mu) \tag{29}$$

where $G_{k-1}(\mu)$ is a polynomial in μ of degree 2^{k-1}. Then:

$$(x_{t+k} = P[P^{k-1}(x_t, \mu)] = [P^{k-1}(x_t, \mu)]^2 + \mu = [x_t^{2^{k-1}} + ... + G_{k-1}(\mu)]^2 + \mu \tag{30}$$

Note that, just as $G_3(\mu) = [G_2(\mu)]^2 + \mu$ in equation (30), the term of the highest degree in x_t is given by $x_t^{2^{k-1} \times 2} = x_t^{2^k}$ and the highest degree of $G_k(\mu)$ is given by the highest degree of $G_{k-1}(\mu)$, that is $2^{k-2} \times 2 = 2^{k-1}$ ∎

Now, we establish the following proposition.

Proposition 4 *Under the conditions of the preceding lemma, the greatest value of the saving rate is $s_{\max} = 1$.*

Proof. To prove the proposition we use the methodology suggested by Myberg [10]. The intuition of the approach is the following. Let us consider the recursive equation $x_{t+1} = f(x_t, a)$, where x is a real variable, a is a real parameter and f is a quadratic map. The values of the bifurcation parameter a can be found by decomposing the successive iterates of the function f into two parts : a polynomial in x and another one in a. Let us note $h_k(a)$ this second polynomial for any integer k. Then, the bifurcation points correspond to the roots of $h_k(a) = 0$ (for a proof of this result the reader is referred to Myberg [10]).

We already know, from the lemma above that the iterates of the map $P(x_t, \mu)$ can be written as the sum of two polynomials, respectively in x and μ. Thus, the greatest value of μ is a solution of $G_k(\mu) = 0$. From the proof of lemma 1 we know that this relation can also be written as:

$$[G_{k-1}(\mu)]^2 + \mu = 0 \tag{31}$$

from which we deduce that:

$$G_{k-1}(\mu) = \pm(-\mu)^{1/2}, \ \mu < 0 \tag{32}$$

Assuming that all the terms of the sequence $(G_k(\mu))$, $k = 1, 2, ...$are positive, the expression above reduces to:

$$G_{k-1}(\mu) = (-\mu)^{1/2}, \ \mu < 0 \tag{33}$$

This last equation can be rewritten:

$$[G_{k-2}(\mu)]^2 + \mu = (-\mu)^{1/2} \tag{34}$$

from which:

$$G_{k-2}(\mu) = [-\mu + (-\mu)^{1/2}]^{1/2} \tag{35}$$

One finally obtains:

$$[-\mu + [-\mu + ... + [-\mu]^{1/2}]^{1/2}]^{1/2} \tag{36}$$

or:

$$-\mu = [-\mu + q]^{1/2}, \ q = [-\mu + ... + [-\mu]^{1/2}]^{1/2} \tag{37}$$

q is the limit of a series antecedents of rank k, defined at $x_t = 0$. Indeed, we have:

$$P^{-1}(0, \mu) = (-\mu)^{1/2}, \ [P^2(0, \mu)]^{-1} = [-\mu + (-\mu)^{1/2}]^{1/2}, etc... \qquad (38)$$

One can easily show that a limit value of this sequence is:

$$\tilde{N}(\mu) = 1/2 + [1/4 - \mu]^{1/2} \qquad (39)$$

Assuming that $\tilde{N}(\mu) = q$, we deduce that the highest value is $\mu = -2$. The equality $B = sA = 1 + [1 - 4\mu]^{1/2}$ therefore implies that $s_{max} = 4/A$. The assumption $A = 4$ implies $s \leq 1$. ■

From these propositions, one concludes that the study of the bifurcations of the map $P(x_t, \mu)$ is possible within the range $(1/4) \leq s \leq 1$.

The approach above can still be used to reformulate some known results from bifurcation theory. More specifically, the functions $G_k(\mu)$, $k = 1, 2, 3, ...$ are needed to show the existence of a period-doubling bifurcation: cycles of doubling periods appear as the saving rate parameter is modified. Further, there also exists a period-halving bifurcation, that is a reduction by half of the periodicity of cycles. This second phenomenon is important because it helps understanding why the fluctuations usually said to be chaotic are not without a "certain order". This "order" reflects a very high sensitivity of the model to the values of its parameters: there are as many attractors as there are possible values for the saving rate when this parameter varies in the interval within which a period-halving bifurcation occurs.

3.3 PERIOD-DOUBLING AND PERIOD-HALVING BIFURCATIONS

In this section we use the approach presented in the proof of lemma 1 to compute the bifurcation points of the saving rate parameter. Following Myberg's approach, we know that those points are simply the roots of the polynomials in μ obtained by computing the iterates of the map $P(x_t, \mu)$.

Period-doubling bifurcation

General formulation

From the proof of lemma 1 we know that

$$G_k(\mu) = [G_{k-1}(\mu)]^2 + \mu \qquad (40)$$

This relation holds for any k, especially when $k = 2^m$, $m = 1, 2, 3, ...$
We are looking for values of μ which satisfy the equality $G_k(\mu) = 0$ with
$G_1(\mu) = \mu$.

Example

If, k=2, then the equality $G_2(\mu) = 0$ implies that $\mu^2 + \mu = 0$ and so
$\mu = 0$ or $\mu = -1$. This equality also implies $-\mu = (-\mu)^{1/2}$ and thus both
solutions can be considered as the fixed points of the first-order difference
equation $-\mu_{i+1} = (-\mu_i)^{1/2}$. We only use the first solution, which implies
$s = (1/A[1 + \sqrt{5}]$.

Results and comments

Repeating this for all subsequent values of k leads to the usual pitchfork
bifurcation, which is characterized by a limit value s2 towards which all the
successive values of the saving rate converge. Table 1 indicates the order of
the stable limit-cycles corresponding to the values of the saving rate. The
cumulation point is s=s2=0.7246620535.

k	2^1	2^2	2^3	2^4
s	0.6035533835	0.70019511271	0.7194537996	0.7235359939
k	2^5	2^6	2^7	2^8
s	0.7244083315	0.7245950748	0.7246350677	0.7246436295
k	2^9	2^{10}	...	2^∞
s	0.7246454651	0.7246458627	...	0.7246620535

Table 1. Saving rate values and period-doubling bifurcation

Period-halving bifurcation

General formulation

The values of the saving rate which bring such a bifurcation are obtained
by seeking the values of μ for which the 2^{k-1} maximums and minimums
of the kth iterates of the map $P(x_t, \mu)$ become identical to the 2^{k-1} fixed
points of its $(k-1)^{th}$ iterates. The procedure requires three steps.

First step

One writes down the equations which give the abscissa of the fixed point of the kth iterate of the map $P(x_t, \mu)$. These points correspond to the 2^k fixed points of a k-period cycle:

$$x_t = P^k(x_t, \mu) \qquad (41)$$

Second step

It is interesting then to note that the iterate of the map $P(x_t, \mu)$ admits 2^{k-1} critical points (that is 2^{k-1} points for which the first derivative of $P(x_t, \mu)$ equals zero). Noting their respective ordinates as $x_c, x_{c_1}, x_{c_2}, ...,$ one assumes that :

$$x_c = \mu, \ x_{c_1} = \mu^2 + \mu, \ x_{c_2} = (\mu^2 + \mu)^2 + \mu, \ ... \qquad (42)$$

or:

$$x_c = G_1(\mu), \ x_{c_1} = G_2(\mu), ..., \ x_{c_{k+1}} = G_{c_{k+2}}(\mu) \qquad (43)$$

Third step

Finally, substituting $x_{c_{k+2}}(\mu) = G_{c_{k+2}}(\mu)$ for x_t in equation (41), one gets:

$$P^k[G_{c_{k+2}}(\mu), \mu] = G_{c_{k+2}}(\mu) \Rightarrow H(\mu) = 0 \qquad (44)$$

For each k, there is one value amongst all the values of μ that verify equation (44), which leads to a stable cycle with an even periodicity.

Example

If, we suppose that x_a is a point of a two-period cycle, then we have $x_a = P^2(x_a, \mu)$ (**step 1**). Assuming that x_a is also a fixed point of $P(x_t, \mu)$, we write $x_a = P(x_a, \mu)$. Thus, by substituting in the left-hand side of the first equation x_a with its expression given in the last equation, one gets $P(x_a, \mu) = P^2(x_a, \mu)$. This equality holds for $x_a = 0.866120202$. Assuming that x_a can be written as $\mu_a^2 + \mu_a$ (**second step**), we substitute this value in the equation $P(x_a, \mu) = P^2(x_a, \mu)$ (**third step**) and get:

$$(\mu_a^2 + \mu_a)^2 + \mu_a = [(\mu_a^2 + \mu_a)^2 + \mu_a]^2 + \mu_a \Rightarrow \mu_a = 0.7246458627 \quad (45)$$

Results and comments

In applying these three steps, one obtains the following values for the saving rate (see table 2), to which correspond stable cycles whose periodicity is decreased by half as one increases the parameter s.

k	2^∞	...	2^{12}	2^{11}
s	0.7246620535	...	0.7249952526	0.726289064
k	...	2^8	2^7	...
s	...	0.7322752378	0.7608547029	...
k	2^4	2^3	2^2	2^1
s	0.7246350696	0.7246436313	0.7246454656	0.7246458627

Table 2. Saving rate values and period-halving bifurcation

Theses results enable us to draw the following conclusions.

(1) One notices that the periodicity of cycles associated with the different values of the saving rate is reduced by half as the value of the parameter is increased. This is why one speaks of a "period- halving" bifurcation, by analogy with the "period-doubling" bifurcation where the periodicity of cycles endlessly doubles. For example, when s=0.7249952526, a stable limit-cycle of order $k = 2^{12}$ coexists with an infinitely number of unstable limit-cycles of order higher than 2^{12}. When s=0.726289064, then the $2^{12^{th}}$-order limit-cycle becomes unstable, and a stable limit-cycle of order $2^{12}/2$ is produced. And so on.

(2) $\hat{s} = 0.7246620535$ appears as a limit "to the right" of the successive values of the saving rate (that is $s < s_2$), just as it appeared as a limit "to the left" for the period-doubling bifurcation (that is $s_2 < s$). It is noteworthy that the existence of a period-halving bifurcation is no way incompatible with the fact that aperiodic fluctuations can take place. Indeed, when s takes the values given in table 2, stable limit-cycles of low-order exist but they are viewed as being "noisy". Such stable periodic cycles may be impossible to determine by visual inspection because they overlap with high-order limit-cycles that are unstable. Suppose for example that the saving rate s is 0.7246454656. We know from table 2 that in this case a limit-cycle of order 2^4 exists and is stable. But there also exists an infinitely number of cycles of order higher than 2^4 which are unstable. So, unless the initial values of the capital-labor ratio are located on the stable attractor, the sequence (k_t) can switch, during a long time, from trajectories belonging to unstable high-order limit-cycles before eventually converging towards the stable $2^{4^{th}}$ order stable limit cycle. this causes fluctuations to be apparently

aperiodic. In a word, very complicated dynamics corresponding to very long transients may not be distinguished from true chaotic motion, in the region corresponding to a period-halving bifurcation.

(3) The coexistence of both types of bifurcations allows the construction of an ordering of all the bifurcations points.To show this, we need a few notations. First, the bifurcation points of the saving rate which lead to period-doubling cycles are noted (s^{2^i}), $i = 1, 2, 3, ...$ Secondly, the saving rate values for which there exists a period-halving bifurcation are noted $(s^{2^i})^*$, $i = 1, 2, 3, ...$ Thirdly, for each i, Ω_{2^i} will designate the interval $[(s^{2^i}), (s^{2^i})^*]$. In the light of what has been previously indicated, we have the following relation:

$$\Omega_{2^i} \subset \Omega_{2^{i-1}} \subset ... \subset \Omega_{2^1} \tag{46}$$

(4) The procedure used above can be generalized to any r^{th} iterate $P^r(x_t, \mu)$ of the function $P(x_t, \mu)$, where $r = m \times 2^i$, m=3,4,5,..., and i=1,2,3,... Suppose that $(s^{2^i}_{m \times 2^i})$ and $(s^{2^i}_{m \times 2^i})^*$ are bifurcation points of the saving rates which allow to define both period-doubling and period-halving bifurcations (just as (s^{2^i}) and $(s^{2^i})^*$ were bifurcation points for $P(x_t, \mu)$). One interesting point is that we know from bifurcation analysis, that cycles of period $m \times 2^i$ exist for values of the bifurcation parameter belonging to a certain range of values leading to period-halving cycles. Here, this would implies that there exists values $(s^{2^i}_{m \times 2^i})$ and $(s^{2^i}_{m \times 2^i})^*$ which lie in the interval $[(s^{2^i}), (s^{2^i})^*]$ for each i. These values lead to another ordering of the saving rate values:

$$... \subset ... \subset \Omega_{m \times 2^{i+2}} \subset \Omega_{m \times 2^{i+1}} \subset ... \tag{47}$$

We therefore write:

$$(s^{2^i})^* \subset ... \subset ... \subset \Omega_{m \times 2^{i+2}} \subset \Omega_{m \times 2^{i+1}} \subset ... \subset (s^{2^{i+1}})^* \tag{48}$$

The existence of a period-doubling bifurcation thereby implies that aperiodic fluctuations when they exist are not without a "certain order". This "order" reflects a strong sensitivity of the model to the saving rate values.

3.4 CHAOTIC FLUCTUATIONS AND CYCLES WITH ODD PERIODS

The Li and Yorke's [11] theorem used by Day in his paper, which has been generalized by Li *et al.* [12], says that when odd-period cycles exist chaotic fluctuations occur. Without going back on what has already been said on this point in the literature, we just seek to draw the reader's attention on the following point. One can construct for odd-period cycles the same type of structure and ordering of the saving rate parameter, as the ones previously described for cycles with even periodicity. In this view, we consider here the period-3 cycle, which corresponds to the odd-period cycle with the highest order in Sarkovskii's [13] relation. Since the methods are similar to those described above, we avoid technical details and expose directly the main results.

Period-3 cycle implies a period-doubling bifurcation: main results

Let us define the interval $[(s^3), (s^3)^*]$, where (s^3) and $(s^3)^*$ correspond for $P^3(x_t, \mu) = P[P[P(x_t, \mu)]]$ to (s^{2^i}) and $(s^{2^i})^*$ for $P^{2^i}(x_t, \mu)$ in the preceding paragraph. We thus decompose the expression of the successive iterates of the map $P^3(x_t, \mu)$ into two polynomials in x_t and μ, which are respectively noted $\Phi'_k(x_t)$ and $\breve{G}_k(\mu)$. The bifurcation points are solutions of the following equations (the approach is the same as above to obtain the period-doubling bifurcation):

$$\breve{G}_k(\mu) = 0, \ \breve{G}_k(\mu) = [\breve{G}_{k-1}(\mu)]^2 + \mu, \ \breve{G}_k(\mu) = \mu, \ k = 3 \times 2^m, \ m = 1, 2, 3, \ldots \tag{49}$$

We have reported in the table below, the values of the saving rate corresponding to (49).

k	3×2^0	3×2^1	3×2^2	3×2^3
s	0.8101064046	0.8141127724	0.8153023078	0.8156292668
k	3×2^5	3×2^6	...	$3 \times 2^\infty$
s	0.8156655041	0.8156854345	...	0.8156854472

Table 3. Saving rate values and period-3 doubling bifurcation

As one can see, there exists an infinitely number of (3×2^i)-period cycles with periodicity endlessly doubled until one reaches the cumulation point s=0.8156854472.

Period-3 cycle implies a period-halving bifurcation: main results

Similarly, one can apply to the map $P^3(x_t, \mu)$ the arguments held above to define the period-halving bifurcation associated to $P(x_t, \mu)$. We have reported in the table below the different values of the saving rate for which successive stable limit-cycles exist and belong to a period-halving bifurcation. As one can see, s=0.8156854472 is again a cumulation point.

k	$3\times 2^{\infty}$	...	3×2^6	3×2^5
s	0.8156854472	...	0.81568544722	0.8156854726
k	3×2^4	3×2^3	3×2^2	3×2^1
s	0.8156884065	0.8157308581	0.8157556535	0.8161786889

Table 4. Saving rate values and period-3 halving bifurcation

Just as was previously seen for the cycles with even periodicity, we can construct a similar ordering here for the period-3 cycle. This ordering leads to the following relation:

$$\Omega_{3\times 2^i} \subset \Omega_{3\times 2^{i-1}} \subset ... \subset \Omega_{3\times 2^1} \qquad (50)$$

As a conclusion, it is seen that both period-halving and period-doubling bifurcations are "universal". Universality expresses that these phenomenon are independent of the periodicity (either even or odd) of the cycles which are successive stable attractors as the saving rate parameter is modified. This finding might be surprising as compared with the results usually mentioned in the literature, which focus on the property of sensitivity to initial conditions. However, the mechanisms that leads to chaotic dynamics exhibit a strong structural instability of the models under study, and this can be carefully checked by using the procedures described above. Of course, the arguments exposed can be generalize to all odd-period cycles.

4 Concluding remarks

This paper intended to discuss two aspects of Day's original contribution on neoclassical one-sector growth and chaotic dynamics. Our arguments can be summarized in two parts.

First, it is usually argued that Day's model introduces modifications of the usual production function in one-sector growth models, which are not economically motivated. Because of this, one may argue that there is no reason why the monotone behavior of standard neoclassical growth model should be transformed into erratic fluctuations. Further, this may lead to cast some doubts on all the models that are based on the same method-ology. In this note we have tried to show that there may be however a possibility to rationalize the notion of a "pollution effect" by considering the macroeconomic consequences of microeconomic interactions. The economy is composed of many firms, which adjust their production and take their decisions of accumulating capital, according to the decisions taken by their competitors. Their mimetic behavior is due to the fact that firms are individually confronted to imperfect information concerning the level of future demand. We introduce here a formal expression of the way firms coordinate their decisions, and it has been introduced here to "grasp" the role of the "pollution effect" which holds such a decisive place in Day's model. In this context, a chaotic dynamic may reflects the fact that volatile expectations of agents are a dominant cause of the irregularity of macroeconomic fluctuations.

Secondly, we have tried to show that the monotone evolution of the capital-labor ratio which is usually observed in standard one-sector growth models may become complex because Day's model produce a strong structural instability: slightly modifications of the saving rate lead to a qualitatively different behavior of the solutions. An important outcome of this, is that the economy may undergo aperiodic fluctuations corresponding to long transients, and not to true chaos. We have exposed a simple method, which helps to highlight the mains causes of this structural instability, that is the overlapping of period-doubling and period-having bifurcations.

References

[1] Day, R., and Gang Z., (1994), "Infrastructure, Restricted Factor Substitution and Economic Growth", *Journal of Economic Behavior and Organization*, 23, pp.149-166.

[2] Day, R., (1995), "L'existence hors de l'équilibre", *Revue Economique*, 46, pp.1461-1471.

[3] Nishimura, K. and Yano, M., (1995), "Non-linear Dynamics and Chaos in Optimal Growth: An Example", *Econometrica*, 63, pp.981-1001.

[4] Grandmont, J.M., Pintus, P., and, De Vilder R., (1996), "Capital-Labor substitution and Competitive Nonlinear Endogenous Business Cycles", *Cepremap working paper*, forthcoming.

[5] Day, R., (1982), "Irregular Growth Cycles", *American Economic Review*, 72, pp.406-414.

[6] Lorenz, H.W., (1993), *Nonlinear Dynamical Economics and Chaotic Motion*, Springer-Verlag, Heidelberg.

[7] Albin, P., (1987), "Microeconomic Foundations of Cyclical Irregularities or Chaos", Mathematical Social Sciences, 13, 183-214.

[8] Day, R., (1986), "Unscrambling the Concept of Chaos through Thick and Thin: Reply", *Quarterly Journal of Economics*, 101, pp.425-426

[9] Melese, F., and Transue, W., (1986), "Unscrambling Chaos Through Thick and Thin", *Quarterly Journal of Economics*, 101, pp.419-423.

[10] Myberg, P.J., "sur k'itération des polynômes réels quadratiques", *Journal de Mathématiques Pures Appliquées*, 41, 339-351.

[11] Li, T.Y., and, Yorke, J.A., (1975), "Period Three implies Chaos", *American Mathematical Monthly*, 82, pp.985-992.

[12] Li, T.Y., Misiurewicz, M., Pianigiani, G., and Yorke, J.A., (1982), "Odd Chaos", *Physics Letters*, 87A, pp.271-273.

[13] Sarkoskii, A.N., (1964), "Coexistence of Cycles of a Continuous Map of a Line into Itself", *Ukranichkii Matematicheskii Zhurnal*, 16, pp.61-71

Part 2

Business Cycles

Hysteresis Arising from Individual Inertia and Behavorial Heterogeneity

Reiner FRANKE

1 Introduction

Appeal to the concept of hysteresis has become widespread in explanations of a variety of economic phenomena. Two major fields are industrial economics, with special reference to international trade, and labour economics, with special reference to unemployment.[1] Generally, hysteresis is a property of dynamic systems. Hysteresis exists when the long-run or final value of a variable not only depends on the long-run values of exogenous variables but also on the system's previous outcomes. This path-dependence implies that different initial conditions, or different exogenous shocks, may lead to different long-run or final configurations. A handy formulation is that "history matters".

In formal models, and particularly in empirical work, economists have tended to view hysteresis as a special case of linear systems of equations. Thus in discrete-time models hysteresis is constituted by the presence of a unit root. Most convenient are stochastic models that are reducible to one state variable, so that hysteresis can be identified with a random walk. These features are usually achieved by specific assumptions about (perhaps lagged or partial) adjustments of some macroeconomic key variables.[2]

Seeking for a more fundamental cause of hysteresis, another direction of research starts out directly from the micro level. It abandons the concept of the representative agent and considers heterogeneous agents with discontinuous, lumpy adjustments. If there are many small decision units and agents are sufficiently dissimilar, the aggregate variables will nevertheless

---

[1] See Franz (1990) for a brief overview.

[2] The influential papers by Blanchard and Summers (1986, 1987) on the persistence of high unemployment in Europe during the 1980s are only one example of how a random walk is constructed from a few number of simple macroeconomic relationships.

change in a smooth way. Owing to the individual inertia, however, past decisions and how they were distributed within the population of agents may still leave their mark on the present. So the whole history of non-synchronized adjustments could, possibly, affect the paths taken by the macro variables. This point has recently been worked out by Cross (1994) in a formal economic model where, on the basis of certain threshold values, agents choose between two polar types of investment strategies in response to the motions of an exogenous expected payoff variable. The modelling device was in fact carried over from the physical sciences and had only to be translated into economic terms.

The approach provides a useful framework to study the micro-macro relationship and its hysteretic potential. An imperative extension is the introduction of a feedback mechanism from the aggregate outcomes of the agents' decisions back to their expectations about the payoff. This task is the subject of the present contribution. The endogeneization of the expected payoff closes the model and gives rise to the notion of equilibrium, i.e., an infinite number of states of the economy where, if unperturbed, the dynamic process essentially comes to rest. In our discussion these are actually steady state positions, with different rates of growth induced by different compositions of the population. Interesting features of hysteresis derive from the fact that, in general, the composition depends on the history of the economy, while, on the other hand, the system's memory is selective and certain periods of the history may also be forgotten.

On the whole, the model is a deterministic nonlinear dynamic system expressed in discrete time (with a flexible adjustment period that may be arbitrarily short). Due to the non-standard description of the population composition of heterogeneous agents it is no ordinary difference equations system, so that the usual tools of a mathematical treatment do not apply. Important points can still be made by means of geometrical arguments. All statements and conjectures are furthermore substantiated by numerical simulations. Economically, the model is a highly stylized representation of the linkages between the real and financial sector. In the present specification, it incorporates a variety of the multiplier-accelerator mechanism and, with respect to investment behaviour, might be said to have a weak Minskian flavour.[3] It should be added that the paper is exclusively concerned with theoretical issues of hysteresis; there is still a long way to go until this kind of research leads to empirical results.

The remainder of the paper is organized as follows. The next section puts the model into perspective. Together with a general characterization

---

[3]The formulation of the dynamic adjustment equations that close the model involves more feedback effects than would perhaps be necessary to put forward our main points. The advantage of the more elaborate model is that it gives rise to smooth, and in a a sense even realistic time paths — although at the present stage of research we do not wish to overemphasize such a step towards calibration.

of the structural features, it also anticipates the main results of our investigation. The formal elaboration begins with the subsequent sections. Section 3 introduces the basic concept of individual inertia and behavioural heterogeneity. In particular, it describes how agents react to changes in the expected payoff and how exogenous variations in the payoff affect the population composition. This section is a short recapitulation of the framework developed by Cross (1994), though it is occasionally phrased in somewhat different terms with a view to the subsequent extension. Section 4 endogenizes the expected payoff and formulates the full model. Sections 5 and 6 study the hysteresis effects that may then emerge. In Section 5, a set of parameters is assumed which ensures that the economy always converges directly to some growth equilibrium. Oscillatory motions are dealt with in Section 6, where both may happen: dampened oscillations (in the relative magnitudes) or sustained endogenous growth cycles. Section 7 concludes. A number of finer details are relegated to an appendix.

2 Perspective of the Model

A basic feature of hysteresis in closed models is brought out more clearly if we neglect repeated random shocks and consider ordinary deterministic systems of difference equations. They may also be nonlinear. A unit root occurs if they possess a continuum of stationary states, i.e., one eigenvalue of the Jacobian matrices evaluated at these points is unity (in continuous-time systems, the Jacobians have a zero root). If it is supposed that the system is n-dimensional, if all the matrices have exactly one unit root, and the (one-dimensional) equilibrium set is attractive, then each of the equilibria is associated with a stable manifold of dimension $n-1$, which together fill out the whole basin of attraction. Conversely, to every point in this basin, which may be a starting point of the dynamic process, there corresponds one stable manifold and one state of equilibrium to which the economy will converge.[4] Stochastic shocks can thus be seen as carrying the economy from one stable manifold onto another. It goes without saying that the set of equilibria is predetermined, i.e., independent of any initial condition or perturbation. In most applications it is also quite easy to compute.

The situation in the model here advanced is more elaborate. Though the concept of heterogeneous agents and the component of inertia in their behaviour is fairly elementary and the population composition admits a

---

[4]An analysis of such deterministic systems can be found in Giavazzi and Wyplosz (1985) for the linear case or, spelling out the nonlinear subtleties, in Franke (1987, pp. 77–92).

unique description by a finite-dimensional vector v, the problem is that in general the dimension of this vector is not bounded: for each integer number N a composition C of the population can be constructed such that the dimension of the associated vector $v = v(C)$ exceeds N. The changes in the population composition as the dynamic process unfolds may also act on the dimension of v, which may decrease as well as increase. The fact that the composition C can be arbitrarily complicated and the dimension of $v(C)$ arbitrarily large entails that the set of all the growth equilibria that are *a priori* possible has a more intricate, even unwieldy structure than the one-dimensional manifold above. Furthermore, the usual investigations of local stability do not work any longer. So the equilibrium set continues to be predetermined, but this fails to be of much use for the analysis.

The concrete analysis of the model is concerned with two topics, where all statements are illustrated by numerical simulations. In the first stage, it is assumed that the economy has settled down in a steady state and that then a one-time shock arrives. We actually concentrate on shocks to a more or less psychological variable characterizing the financial sector, which is called financial distress. The possible reactions of the economy fall into four categories: the economy may find back to the old equilibrium in a direct way, it may converge to a new equilibrium in a direct way, it may oscillate towards a new equilibrium position, or the oscillations do not die out and the economy approaches a periodic orbit, i.e., a growth cycle. An important finding is that the shocks have *asymmetric effects*. In the case of direct convergence, for example, a small or medium negative shock may increase the growth rate in the resulting new equilibrium, whereas after a positive shock of the same size the economy returns to its original position. Another phenomenon, which occurs under cyclical adjustments, is that both positive and negative shocks induce convergence to new equilibria, and all of them exhibit either a lower or a higher steady state growth rate. Which of the two cases prevails depends on the attributes of the population composition in the initial equilibrium position.

If the system does not approach an equilibrium but enters into persistent fluctuations, the strength of the shock may affect the amplitude of the periodic motions. Generalizing the notion of hysteresis discussed above, which was related to the points of rest of a dynamic system, these effects may be called hysteresis in closed orbits, or *cyclical hysteresis* for short.

As regards the stability of steady states, a phenomenon can emerge that, to our knowledge, has not been observed before in dynamic modelling. The initial equilibrium may be attractive if the shocks to financial distress are sufficiently small, while the economy displays sustained cyclical growth if the shocks become larger. So far, this is a situation which, according to Leijonhufvud's expression, is known as corridor stability. A new peculiarity is that when the shocks are further increased in size, there is another switch in the system's qualitative behaviour and the economy follows a course back to a steady state position again, though growth takes place at a different level. Reserving the term stability to indicate convergence to a single state of equilibrium, it might be said in a pointed manner that medium-sized shocks are detrimental to stability, whereas large shocks are stabilizing.

In a second stage of the analysis, we revert to the issue of a random walk as a reduced form in which hysteresis is often seen to manifest itself. To this end the system is subjected to repeated random shocks at regular intervals. If the economy always converges to an equilibrium, an old one or a new one, and the shocks occur sufficiently infrequently, one can focus attention on the associated growth rates and how they change in a stochastic manner. Formally, the sequence of these steady state growth rates can be viewed as a random walk where, however, the random innovations are no longer independently and identically distributed. We would thus obtain a 'perturbed' random walk, with induced innovations. Nevertheless, if the law of motion underlying the time series of the equilibrium growth rates is not known, the hypothesis of a pure random walk might still be considered a useful approximation. This is, after all, the basis of the broad literature on a unit root in GNP.

By studying the distribution and the time series characteristics of the innovations in the growth rates and comparing them to the original shocks, it is possible to get some information on the goodness of this approximation, whether it proves workable or not. We do not provide an extensive Monte Carlo simulation, but the few sketches of our realizations of the stochastic experiment are sufficient to indicate that the 'perturbations' of the random walk may not be negligible.

A typical feature of a time series generated by a pure random walk is that over some passages of time it hovers around a certain value. Of course, these levels will shift from one such phase to another. In our context one may look upon them as different growth 'regimes'. A less rigorous conception of hysteresis might content itself with a visual inspection of the time series and an assessment of different growth regimes. This point of view needs no theoretical background and, in particular, is applicable whether the economy converges to steady growth paths or produces undampened oscillations. These growth regimes can also be observed in our experiments in the case of point-convergence. In the case of sustained endogenous growth cycles, by contrast, the significance of hysteresis in this sense seems to vanish. Though the shocks modify the amplitudes of the oscillations, the effects do not visibly carry over to the time averages. Accordingly, these economies could even be classified as trend-stationary. We have therefore a theoretical relationship between hysteresis in growth regimes and the (paradigmatic) problem whether the observed cyclical behaviour is basically endogenous or brought about by exogenous shocks to the economy.

3 The Basic Description of Changes in the Population

We distinguish two polar types of strategies adopted by agents. They may be called prudent and bold, where in the present section the corresponding actions can be left unspecified. Agents are not stuck in such one type but decide on the basis of an expected payoff, that is, they change their strategies if the payoff passes certain threshold values. In detail, an individual agent is characterized by two critical values λ and μ of the expected payoff, with $\lambda \leq \mu$. If this agent is presently prudent, a payoff $\pi \geq \mu$ is required to induce her/him to switch to the bold type of behaviour. A switch back from bold to prudent is made when the payoff falls down to, or below, the threshold value λ. If one wishes to invoke optimizing behaviour, the two strategies might be viewed as corresponding to the outcome of an optimal control problem for whose solution, in the presence of certain adjustment costs, the bang-bang principle applies. Examples are two-sided (S, s) rules (e.g., Miller and Orr 1966) or Dixit's (1992, 1995) argument that, under irreversible investment, waiting for more information to arrive has a positive value which leads to discontinuities in the investment decisions. These results may furthermore corroborate discontinuous adjustment rules as a reasonable policy if agents are faced with irreducible uncertainty.

The agent in our setting has a specific range of the payoff π, $\lambda \leq \pi \leq \mu$, inside which it is necessary to know the history of the payoff movements in order to determine which type of strategies is chosen. The switching rules can be illustrated by the rectangular loop in an input-output diagram; see Figure 1 (the payoff π is the input, bold or prudent behaviour is the output). As π is continuously rising, the ascending branch *abcde* is followed, while the descending branch *edfba* is traced when π is continuously decreasing.[5]

---

[5]In the language of the physical sciences, these rules define an elementary hysteresis operator with local memory; cf. Mayergoyz (1991), pp. *xiv*, 2.

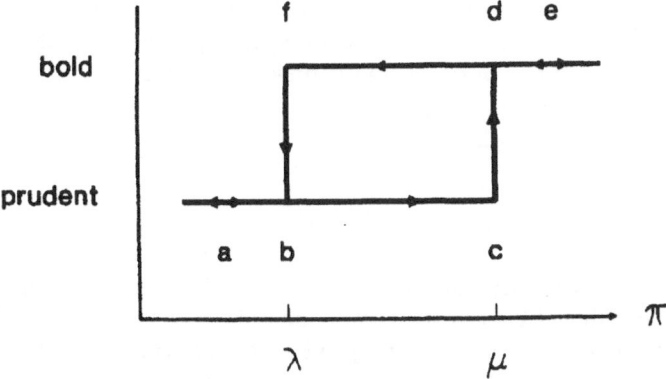

Figure 1 Switching rules of an individual agent

Generally, individual agents have different switching values. Reasons may be different attitudes towards risk, or different technological and institutional characteristics that may limit the flexibility to change strategies. This is, however, the only dimension of behaviour in which agents are supposed to differ. In particular, they share the same expectations about the relevant payoff. Accordingly, behavioural heterogeneity is constituted by the distribution of the pairs of switching values (λ, μ) across the population of agents. If λ and μ are both contained in an interval $[\lambda_{min}, \mu_{max}]$, we have a one-to-one correspondence between individual agents and pairs (λ, μ) in the triangle

$$\Delta := \{(\lambda, \mu) : \lambda_{min} \leq \lambda \leq \mu \leq \mu_{max}\}$$

To ease the exposition, let us work with a continuum of agents and assume that their characteristics (λ, μ) are uniformly distributed on Δ. The latter supposition will also make clear that no special clustering of agents is needed to obtain the hysteresis effects to be investigated below.

As will be seen shortly, at any instant of time the triangle Δ is subdivided into two regions representing the agents who have presently chosen the bold and prudent strategies, respectively. The boundary between the two sets is formed by a staircase line as exemplified in Figure 2. This line, which is called the interface, is defined by finitely many vertices, say there are m of them,

$$V_k = (\lambda_k, \mu_k), \qquad k = 0, 1, \ldots, m$$

They may be ordered as

$(\lambda_o, \mu_o) = (\lambda_{min}, \mu_{max})$ and $\lambda_k > \lambda_{k-1}$, $\mu_k \leq \mu_{k-1}$ for $k = 1, \ldots, m$

There is, however, only one exception, namely for $m = 1$ and $\mu_o = \mu_1 = \mu_{max}$, where an inequality in the μ-coordinates might not be strict. The final stair is given by the vertical line connecting vertex $V_m = (\lambda_m, \mu_m)$ with the hypotenuse of the triangle Δ if $\lambda_m < \mu_m$; otherwise V_m is itself a point on the hypotenuse. The entire interface, i.e. the set of these vertices, is denoted by V,

$$V = \{V_o, V_1, \ldots, V_m\}$$

Agents with the lower switching values, (λ, μ) below the interface, are bold, while switching values (λ, μ) in the complementary region above V indicate that such an agent has opted for the prudent set of strategies. Hence, by virtue of the hypothesis of a uniform distribution of agents, the proportion of bold agents in the total population is simply determined by the area below the staircase line, divided by the total area of the triangle Δ.[6] In explicit terms, let $T_k = T_k(V)$ be the area of the trapezoid defined by the vertices V_{k-1} and V_k. That is, starting from $V_k = (\lambda_k, \mu_k)$ in clockwise orientation ($k = 2$ in Figure 2), T_k is bounded by the four points (λ_k, μ_k), (λ_k, λ_k), $(\lambda_{k-1}, \lambda_{k-1})$, (λ_{k-1}, μ_k), so that

$$T_k = T_k(V) = (\lambda_k - \lambda_{k-1})\left[\mu_k - \lambda_k + (\lambda_k - \lambda_{k-1})/2\right]$$

If the final vertex V_m comes to lie on the hypotenuse of Δ, $\lambda_m = \mu_m$, the trapezoid T_m degenerates to a triangle. Writing $T(\Delta)$ for the area of the limiting triangle Δ itself, one has

$$T(\Delta) = (\mu_{max} - \lambda_{min})/2$$

The share of bold agents in the total population, which we designate b, can therefore be represented by the mapping $B = B(V)$,

$$b = B(V) := \frac{1}{T(\Delta)} \sum_{k=1}^{m} T_k(V) \tag{1}$$

where the interface V is historically given.

---

[6]In the general case, the distribution of agents may be represented by a density (or weight) function on Δ. Calculating the proportion of bold agents then involves the evaluation of an integral, similar as in Mayergoyz (1991), p. 2.

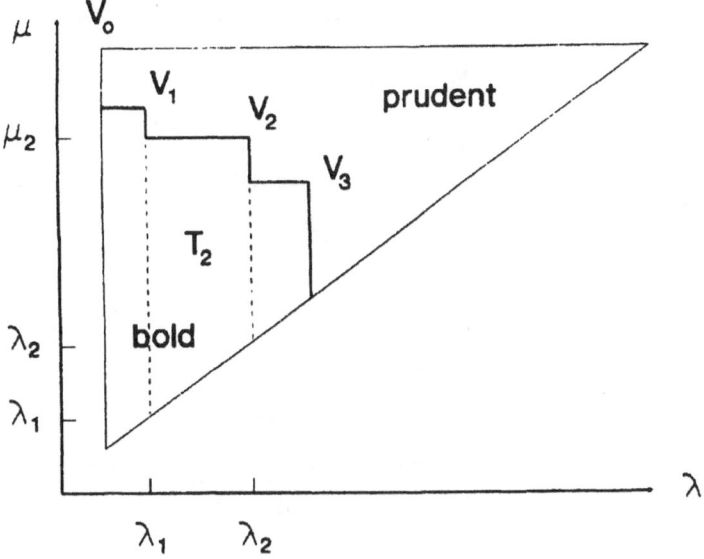

Figure 2 The interface $V = \{V_0, V_1, V_2, V_3\}$

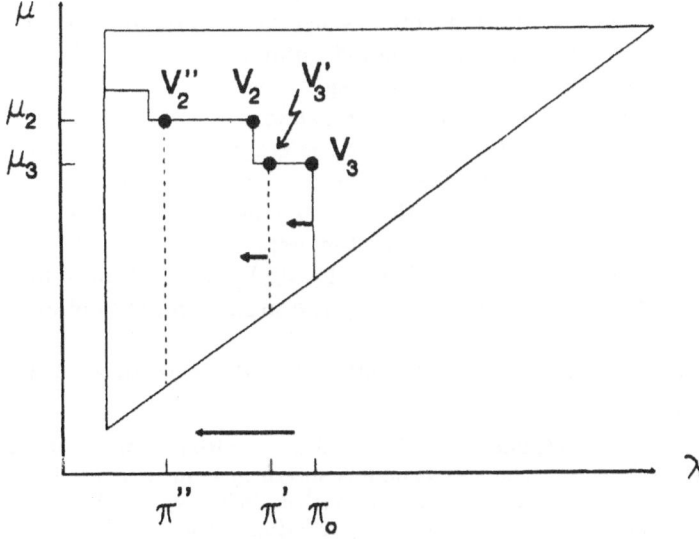

Figure 3 A decrease in th payoff down do π'', wiping out vertex V_2

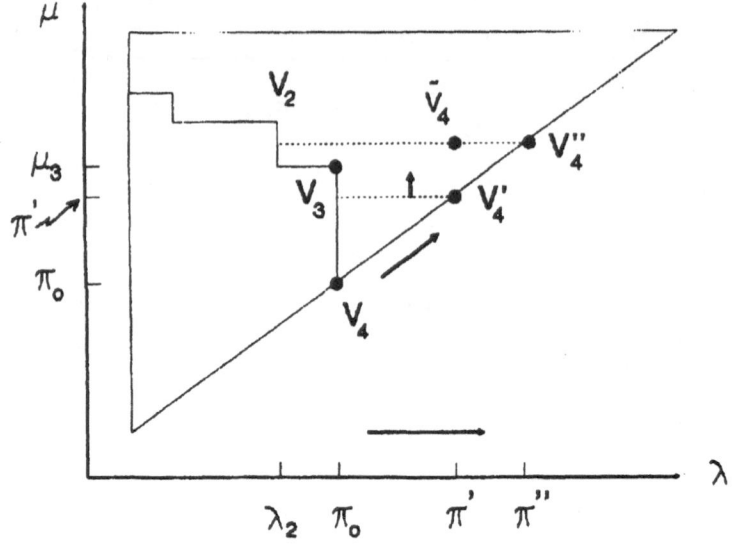

Figure 4 An increase in the payoff up to π'', wiping out vertex V_3

To start the discussion of how variations in the payoff π affect the population composition of bold and prudent agents, consider Figure 3. The interface $V = \{V_o, V_1, V_2, V_3\}$ corresponds to the present payoff $\pi = \pi_o$. Now, let π decrease to π', which should diminish the share of bold agents. In fact, this change induces the bold agents with a 'downward' switching value λ between π_0 and π' to return to prudent behaviour. Their 'upward' switching value μ is less than μ_3. Geometrically, these are the agents with switching values (λ, μ) contained in the trapezoid bounded by V_3' and V_3. In other words, the decline in the payoff shifts the vertical line to the left which links the vertex V_3 and the hypotenuse of Δ. So the inital interface V is modified to $V' = \{V_o, V_1, V_2, V_3'\}$. The (normalized) area below V' represents the corresponding new proportion of bold agents, i.e., $b = B(V')$.[7]

Next, suppose that the payoff continues to decrease to $\pi'' < \pi'$, which

---

[7]Of course, the argument equally applies if initially all agents are bold because of $\pi_o \geq \mu_{max}$. The corresponding interface V in this case is just given by the top line of the triangle Δ, the vertices being $V_o = (\lambda_{min}, \mu_{max})$ and $V_1 = (\mu_{max}, \mu_{max})$; cf. the remark on the ordering of the vertices.

moves the vertex V_3' further to the left. At the moment, however, when the payoff reaches λ_2, where λ_2 corresponds to the vertex V_2, there is a sudden disproportional increase in the number of switching agents. Here we also encounter bold agents with an upward switching value $\mu \geq \mu_3$ (but $\mu \leq \mu_2$), who now switch in their behaviour back to prudent. The geometric implication is that as π resumes its fall below λ_2, the previous vertex V_2 is *wiped out*. From then on it is the vertical line connecting V_2 with the hypotenuse that shifts to the left, until $\pi = \pi''$. Writing $V_2'' = (\lambda_2'', \mu_2'') = (\pi'', \mu_2)$, the change in π from π' to π'' shapes the new three-vertex interface $V'' = \{V_o, V_1, V_2''\}$. The elimination of vertices will later be an important feature in the analysis of hysteresis effects.

In describing the effects of an increase in the payoff from the initial value π_o, as sketched in Figure 4, a new vertex emerges. As a first step, define V_4 as the point $V_4 = (\lambda_4, \mu_4) = (\pi_o, \pi_o)$. Then as π starts to rise, V_4 moves upwards along the hypotenuse of Δ. At $\pi = \pi'$, say, prudent agents with an upward switching value μ between π_o and π' have converted to boldness. These are the agents with switching values (λ, μ) in the triangle specified by the corners (π_o, π_o), (π_o, π') and (π', π'). Setting $V_4' = (\pi', \pi')$, the interface corresponding to the payoff $\pi = \pi'$ is given by $V' = \{V_o, V_1, V_2, V_3, V_4'\}$.

Continuing the increase in π shifts the horizontal line with end-points V_4' and $(\pi_o, \pi') = (\lambda_3, \pi')$ upwards (note that $V_3 = (\lambda_3, \mu_3) = (\pi_o, \mu_3)$). As π rises beyond μ_3 we observe a similar effect as above, namely, the vertex V_3 is wiped out. There are also prudent agents with a downward switching value λ less than $\pi_o = \lambda_3$ (but exceeding λ_2) who adopt the bold strategies, if $\mu_3 \leq \mu \leq \pi$ with respect to their upward switching value μ. The payoff $\pi = \pi''$ therefore gives rise to the interface $V'' = \{V_o, V_1, V_2, V_4''\}$, where $V_4'' = (\pi'', \pi'')$. The vertices in the set V'' would, of course, be renumbered before the analysis were taken any further.

The rules determining the modification of the interface as the payoff π changes may be compactly summarized by a mapping $\Phi = \Phi(\pi, V)$,

$$\Phi(\pi, V) = \text{interface } V' \text{ resulting from (the old) interface } V \quad (2)$$
$$\text{and (the new) payoff } \pi$$

The terms old and new can be seen as referring to the previous and the present period in a discrete-time setting. It is then understood that the old payoff π_o, in the previous period, constituted the λ-coordinate of the final vertex V_m, that is, $\lambda_m = \pi_o$. The precise definition of the mapping Φ in algebraic terms is given in the Appendix.

We are now also in a position to reconstruct the motions of the payoff that may have generated the interface $V = \{V_o, V_1, V_2, V_3\}$ in Figure 2. Initially, π may have fallen short of λ_{min} and all agents may have been prudent. A steady rise in π until $\pi = \mu_1$ and then down to $\pi = \lambda_1$ will have produced vertex V_1. Vertices V_2 and V_3 are the outcome of the subsequent oscillations in π with peaks μ_2 and μ_3 and the interposed trough λ_2. The payoff is then presently at $\pi = \lambda_3$.[8] Note that the speed of these payoff variations has no bearing on the shape of the interface.

To sum up the impact of fluctuations in the payoff π on the population composition, the vertices of the interface V are brought into existence by the turning points of these motions. Furthermore, each peak in π wipes out the vertices of V whose μ-coordinates are below this local maximum of the time series, while each trough wipes out the vertices whose λ-coordinates are above this local minimum.

After this introduction into the working of the model, a basic hysteresis effect can be revealed. The aggregate behaviour of agents in macroeconomic theory is usually modelled in a way that assigns to every vector of independent variables a uniquely determined outcome of the individual actions. For example, the theory assigns to each set of prices a definite quantity of a good demanded or supplied in the aggregate. If the independent variable is the expected payoff π, then according to this supposition every value of π would be associated with a specific value b of the share of bold agents in the total population (to which might correspond a certain level of aggregate demand for physical capital goods or a financial asset). However, in the present framework with the discontinuous adjustments of the microeconomic units such a functional relationship holds no longer true.

As a case in point, refer to Figure 4 and consider the share of bold agents b that may be induced by the payoff π'. As discussed above for the upward movement of the payoff, b was given by the (normalized) area below the interface $\{V_o, V_1, V_2, V_3, V_4'\}$. Now, suppose that the payoff continues to rise until π'' and subsequently returns to the previous value π'. Owing to the inertia in the behaviour of the individual agents, not all agents having switched from prudent to bold in the upturn of π have also switched back to prudent during the subsequent descent. Geometrically, the agents who have turned bold in response to the rise of π from π' to π'' are represented by the area between the two dashed lines in Figure 4, which is bounded by the lines connecting the six points $V_4' = (\pi', \pi')$, $V_4'' = (\pi'', \pi'')$, (λ_2, π''), (λ_2, μ_3), $V_3 = (\lambda_3, \mu_3)$, and (λ_3, π'). From these bold agents only those with switching values (λ, μ) in the triangle V_4', $\tilde{V}_4 = (\pi', \pi'')$, V_4'' have resumed the prudent behaviour as the payoff has fallen back to π'. This argument shows that in order to determine the share of agents who, given the expected payoff π, have adopted the set of bold strategies, one must know the history of the time path of π. To be precise, not the history in every detail but only the turning points of the payoff, and only those that are not dominated by subsequent turning points, i.e., those that have not been wiped out in the course of time.

---

[8]Alternatively, V may be reconstructed from a 100% bold population and an initial payoff $\pi \geq \mu_{max}$.

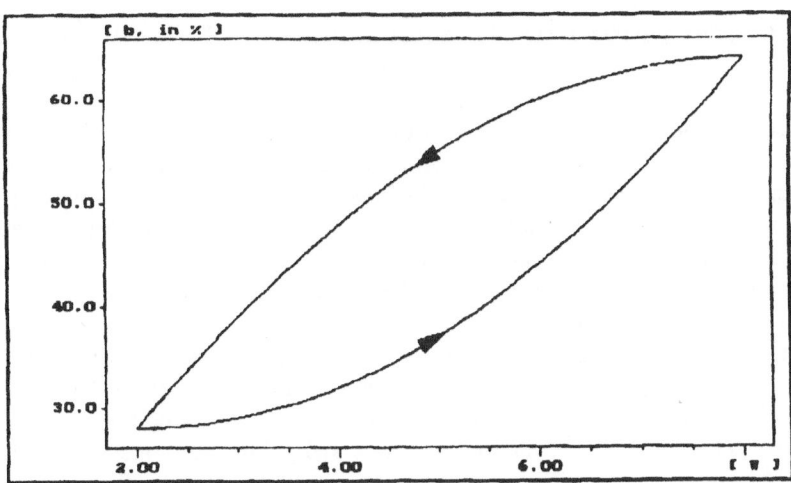

Figure 5 Hysteresis loops arising from a periodic motion of the payoff π (b the share of bold agents)

We illustrate the hysteresis phenomenon in Figure 5. It is another input-output diagram, but unlike Figure 1 the output is here a continuous variable, the share of bold agents b. Underlying are strictly periodic cycles of the payoff.[9] The loop in Figure 5 is typical for the visualization of a hysteresis operator and its branch-to-branch transition after input extrema (see, e.g., the introductory section of Mayergoyz 1991). Additional branches would be generated if the periodic pattern of the payoff were relaxed and π exhibited different turning points. It may also be noted that information about the present output b and all future values of the input π are not sufficient to deduce the future evolution of b in a unique way. A given pair $b = b(t_o)$ and $\pi = \pi(t_o)$ at time t_o may be compatible with different histories of the payoff, so that, depending on the shape of the interface thus brought about, the same time path of π from t_o on may generate quite different time paths of the population composition.[10]

---

[9]The trough and peak values of π being given by π' and π'', respectively, the interface at the trough value was constituted by $\{(\lambda_o, \mu_o), (\pi', \pi'')\}$, and at the peak value by $\{(\lambda_o, \mu_o), (\pi'', \pi'')\}$. With π on the downswing, the interface was $\{(\lambda_o, \mu_o), (\pi, \pi'')\}$, on the upswing it was $\{(\lambda_o, \mu_o), (\pi', \pi''), (\pi, \pi)\}$.

[10]In the terms of the physical sciences, we have a hysteresis transducer with nonlocal memory, constructed as a superposition of elementary hysteresis nonlinearities with local memories (cf. Mayergoyz 1991, pp. $xiv - xvi$, 3).

4 A Complete Model with Endogenous Payoffs

In the previous section, the evolution of the expected payoff was treated as exogenously given. We now extend the basic framework and introduce a number of new variables and relationships such that the payoff π is fully endogenized. The resulting dynamic model, if unperturbed, is able to produce steady state growth paths as well as oscillatory motions. The subsequent analysis is concerned with their potential to give rise to additional hysteresis phenomena.

In the following the microeconomic units are conceived as heterogeneous firms, which generally expand at different speeds. Since part of their fixed capital investment must be externally financed, these firms are also faced with the choice of adopting more or less risky financial postures. Let a firm's financial practices be represented by a single variable called *exposure*. A higher value of a firm's exposure may, for example, be associated with higher borrowing relative to cash flow, or a greater proportion of short-term debt in total liabilities. Limiting firms to two types of behaviour in this respect, which we continue to identify as bold and prudent, the corresponding levels of exposure are denoted by e_b and e_p $(e_b > e_p)$. They are coupled with two distinct growth rates of the stock of physical capital, g_b and g_p $(g_b > g_p)$, to which the bold and prudent firms respectively aspire. The switching of individual firms between the two types of policy is governed by the same principle as above.

A higher value of exposure indicates greater vulnerability to disruptions in the normal functioning of the financial sector. The financial market conditions are captured by a measure of *distress*, d. High levels of distress can be related to high interest rates and an unwillingness of banks to refinance positions as the firms' debt payments become due. Low values of d, on the other hand, are reflective of a state of tranquillity. Certainly, d has also a psychological dimension. Financial distress will be a key determinant of the expected payoffs to the two strategies of exposure.

Besides the differential payoff π and the interface V in the (λ, μ)-plane, other variables in our economy are e, aggregate exposure; g^k, the growth rate of the aggregate capital stock; g^i, the growth rate of aggregate (net) investment; and g, the growth rate of aggregate output which the firms expect to prevail on average over the next few months or even years. With a view to the ensuing computer simulations, the formulation is in discrete time with a fixed adjustment period of length h.[11] At the beginning of the present period t, the variables $\pi_{t-h}, V_{t-h}, g^k_{t-h}, g_{t-h}, d_{t-h}$ are historically

---

[11]Let us say that time is measured in 'years'. The process of going to the limit, $h \to 0$, will mathematically be well-defined and would result in a piecewise specified system of differential equations. That is, a 'regime switching' would occur if a vertex of the interface is wiped out.

given from the previous period. We first state the recursive adjustment equations to determine, on this basis, the state of the economy in period t, subsequently a few comments are added (further details can be found in the appendix). With respect to positive constant adjustment parameters $\beta_k, \beta_g, \beta_d$ and three functions $f_y = f_y(g^i)$, $f_d = f_d(e, g)$, $f_\pi = f_\pi(d, g)$, the dynamic system reads as follows:

$$V_t = \Phi(\pi_{t-h}, V_{t-h}) \tag{3}$$

$$b_t = B(V_t) \tag{4}$$

$$g_t^k = g_{t-h}^k + h\,\beta_k\,[b_t\,g_b + (1 - b_t)\,g_p - g_{t-h}^k] \tag{5}$$

$$g_t^i = g_{t-h}^k + (1 + hg_{t-h}^k)\,(g_t^k - g_{t-h}^k)\,/\,hg_{t-h}^k \tag{6}$$

$$g_t = g_{t-h} + h\,\beta_g\,[f_y(g_t^i) - g_{t-h}] \tag{7}$$

$$e_t = b_t\,e_b + (1 - b_t)\,e_p \tag{8}$$

$$d_t = d_{t-h} + h\,\beta_d\,[f_d(e_t, g_t) - d_{t-h}] \tag{9}$$

$$\pi_t = f_\pi(d_t, g_t) \tag{10}$$

Equations (3) and (4), which determine the new interface and the resulting share of bold firms, are known from Section 4. In the aggregate, firms intend to expand their capital stocks at the rate $b_t g_b + (1-b_t)g_p$. Equation (5) acknowledges the financial and technical difficulties of adjusting actual to desired capital stocks and supposes a partial adjustment mechanism towards this target rate with coefficient β_k. Equation (6) is no more than an identity for the growth rate of investment, the discrete-time analogue of the equation $\hat{I} = \hat{K} + \hat{g}^k$ in continuous time (which is obtained from the logarithmic differentiation of the definition $g^k = \hat{K} = I/K$, where I is net investment, K the capital stock, and the caret designates growth rates).

The term f_y in equation (7) stands for the actual growth rate of aggregate output. Based on the notion of a multiplier process, which moderates in a depression or boom phase, f_y is devised as an increasing S-shaped function of the growth rate of investment (under 'normal' conditions, i.e. over a medium range of g^i, the two growth rates of output and investment coincide). As firms are supposed to live in an uncertain environment and g is to represent output expectations, not for the next adjustment period, but over a longer time horizon, we stipulate that g is revised only partially in the direction of the currently observed growth rate f_y. The emerging equation (7) is formally equivalent to the specification of adaptive expectations.[12]

Equation (8) defines aggregate exposure. In the determination of the

---

[12] Arguments from the literature which defend the use of an adaptive expectations mechanism in macroeconomic modelling are collected in Franke (1994). Suffice it here to mention the contribution by Heiner (1988, especially p. 272), who demonstrated that in an uncertain world with imperfect information, partial adjustments indeed represent a reasonable mechanism of the revision of expectations.

changes in financial distress d, note first that distress on the credit markets depends on the (expected) cash flows of firms as well as their amount of contractual payments. These aspects are here captured by the expected macroeconomic growth rate g and aggregate exposure e. The function $f_d = f_d(e, g)$ in (9) represents the corresponding *self-sustaining* level of distress; the notion is that this level would be reproduced over time if e and g remained constant. Clearly, $\partial f_d/\partial e > 0$, $\partial f_d/\partial g < 0$. However, we assume that it takes some time until the changes in e and g take full effect in new loan contracts. Accordingly, equation (9) describes gradual adjustments of the actual degree of financial distress towards its self-sustaining level f_d, with adjustment speed β_d.

Finally, equation (10) specifies the payoff expectations as being influenced by the conditions firms find on the financial markets, and by the prospects of future economic growth. The expected differential payoffs π of bold strategies with their higher indebtedness are therefore negatively affected by financial distress, while the more aggressive policy will yield higher increases in profits than the prudent strategies when the economy expands at a faster pace. Hence $\partial f_\pi/\partial d < 0$ and $\partial f_\pi/\partial g > 0$.

Although equations $(3-10)$ constitute a well-defined dynamic process and the computations involved are straightforward, there is an important point in which they differ from ordinary systems of difference equations. The latter exhibit a definite number of state variables, which is the dimension of the system. Because of the central role of the interface, it is here impossible to determine a (finite) dimension. The interfaces may be parametrized by their vertices, so that the dimension of a given interface would be given by twice the number of its vertices. However, if the process generates cyclical motions then, as seen in the preceding section, the number of vertices of the interface may increase. The problem is that we cannot tell in advance the number of vertices to be shaped in the course of the process; it might even be infinitely many.

System $(3-10)$ gives rise to a notion of equilibrium. In a rest point, the four growth rates will all be equal, $g = g^k = g^i = f_y(g^i)$. Expressing their relation to the share of bold firms as $g^* = g^*(b) := b\, g_b + (1 - b)\, g_p$, and similarly for exposure in (8), $e = e(b)$, the corresponding level of financial distress is obtained from (9) as $d = f_d^*(b) := f_d(e(b), g^*(b))$. Inserting it in function (10) yields the payoff

$$\pi^* = \pi^*(b) := f_\pi[\, f_d^*(b), g^*(b)\,] \tag{11}$$

On the other hand, equations (3) and (4) have to be taken into account. If we now have a given interface V, this means that *via* the payoff π^* and the mapping Φ, the newly generated interface V^* must be reproduced in the equilibrium. That is, $\Phi(\pi^*(b), V^*) = V^* = \Phi(\pi^*(b), V)$. V^* is reproduced if (and only if) the payoff does not change or, what amounts to the same, if the share of bold firms remains constant. A state of rest of the economy is therefore characterized by a solution of the following fixed-point equation in b,

$$b \;=\; B[\,\Phi(\pi^*(b), V)\,] \tag{12}$$

It can be shown that π^* is a decreasing function of b if the growth rates effects in the distress and payoff functions in (9) and (10) are of moderate size (i.e., if the partial derivatives $|\partial f_d/\partial g|$ and $\partial f_\pi/\partial g$ are not too large), a condition that will be met for all sets of numerical parameters considered in this paper. From the discussion of Figures 3 and 4 we know that the area $B = B[\Phi(\pi, V)]$ decreases as π decreases. b furthermore exceeds (falls short of) the right-hand side of (12) if $b=1$ (if $b=0$). Hence there is exactly one value of b satisfying equation (12). To sum up, for every interface V there exists another, uniquely determined interface V^* supporting an equilibrium composition of the population of firms. We may succintly write

$$b^* \;=\; b^*(V) \;, \qquad \begin{array}{l} \text{equilibrium share of bold firms,} \\ \text{with respect to an initially given interface } V \end{array} \tag{13}$$

Since the given interface V may have any shape, this argument demonstrates that there exists a huge set of steady state positions, much 'larger' indeed than a one-dimensional manifold. The interface $V^* = \Phi(\pi^*(b^*), V)$ itself can be viewed as constituting a reference equilibrium for the actual economy in a state with interface V. V^* will generally be shifted as V changes in the course of the dynamic process.

5 Hysteresis Phenomena in the Closed Model

The presence of multiple equilibria has a bearing on the notion of stability. While in a deterministic system with a unique equilibrium, (asymptotic) stability means it is this specific state that attracts the trajectories, it can now only be asked whether the economy will converge to *some* state of equilibrium. Here it is the dynamic system as such that will be called stable, with respect to a given initial condition. If convergence obtains for all initial conditions, we will call the system *universally stable*. Specifying an initial condition, which includes an initial interface V, it would nevertheless be extremely hard to determine the equilibrium in advance in which the economy will finally settle down, even more so when the associated equilibrium share of bold firms, $b^* = b^*(V)$ in terms of equation (13), varies over time. Apart from that, it will be seen below that the system might be stable only for a subset of starting positions.

Given an equilibrium interface with payoff π^*, which connects at $(\lambda, \mu) = (\pi^*, \pi^*)$ to the hypotenuse of the triangle Δ, and an initial value π_o of the payoff close to π^*, a comparison of the trapezoids in Figures 3 and 4 shows that the calculation of the area below the new interface brought about by π_o uses different expressions in equation (1), depending on whether $\pi_o > \pi^*$ or $\pi_o < \pi^*$. Thus locally around every equilibrium interface there are two distinct regimes of difference equations. This property of the system precludes an application of the standard tools of stability analysis. Besides some basic geometrical arguments in the (λ, μ)-plane which help to classify the main cases that may occur, we therefore have to resort to numerical investigations.

In the following, we confine our interest to two parameters and their impact on stability. They refer to the distress and the payoff function in (9) and (10), which we now specify as

$$f_d(e, g) = \min\{1, \max\{0, \frac{e - e_p}{e_b - e_p} - \alpha_d g + \gamma_d\}\} \tag{14}$$

$$f_\pi(d, g) = \sigma_\pi \cdot [(1 - d)(1 + \alpha_\pi g) - d\psi] + \gamma_\pi \tag{15}$$

($\alpha_d, \alpha_\pi, \gamma_d, \gamma_\pi, \sigma_\pi, \psi$ are all constant parameters.) Function (14) limits distress to values between zero and one, so that in the discussion below, d and the shocks to d can be expressed in percentage terms. The intercept γ_d in (14) serves to shift the initial equilibrium value of d to a convenient level. Similarly, σ_π and γ_π scale and shift the payoff function (15) such that the model's initially given interface can bring about a particularly nice equilibrium composition of firms, namely, a 50 per cent share of bold firms, b_o^*. The system always starts from this equilibrium configuration, which is then disturbed by an exogenous shock to d.

The parameter ψ in (15) represents the differential distress losses of bold firms if d is high. The effects of expected output growth, which reduce financial distress and raise the payoffs to the bold strategies, are measured by α_d and α_π, respectively. It is these latter coefficients on which we concentrate, where three pairs (α_π, α_d) are considered. Somewhat loosely, they may be characterized as follows.

Scenario 1: no growth rate effects ($\alpha_\pi = 0$, $\alpha_d = 0$); universal stability, where the (new) equilibrium is approached monotonically.

Scenario 2: moderate growth rate effects ($\alpha_\pi = 4$, $\alpha_d = 1$); universal stability, where the (new) equilibrium is approached in a cyclical manner.

Scenario 3: strong growth rate effects ($\alpha_\pi = 5$, $\alpha_d = 2.5$); no convergence, but persistent growth cycles (for medium-sized shocks at least).

We are mostly concerned with the stable scenarios, as they can be employed to investigate a possible relationship of system (3–10) to a random walk process. The economy is correspondingly subjected to repeated shocks.

In the present framework, the variables most prone to such exogenous perturbations are associated with the financial markets and their psychology, which is our reason for focusing attention on shocks to the level of financial distress, d. It is furthermore assumed that these shocks arrrive infrequently. That is, after the economy has been thrown out of a state of rest by a shock to d, it has sufficient time to converge close to a — possibly new — equilibrium until the next shock occurs. It may be conjectured that if the shocks are i.i.d. with zero mean, then, though perhaps somewhat distorted, this stochastic process has the basic features of a random walk.

To scrutinize this idea, let the economy start from an initial equilibrium with a share of bold firms $b_o^\star$ and consider the sequence $b_1^\star, b_2^\star, b_3^\star, \ldots$ of equilibrium shares of bold firms that is generated by a sequence of random shocks to financial distress. The change from one equilibrium $b_k^\star$ to the next $b_{k+1}^\star$ may be summarized by a function $F = F(\delta_k, V^\star(k))$, where δ_k denotes the shocks to d and $V^\star(k)$ is the equilibrium interface associated with $b_k^\star$. Thus,

$$b_{k+1}^\star \;=\; b_k^\star \;+\; F(\delta_k, V^\star(k))\,, \qquad\qquad k = 0, 1, 2, \ldots \qquad (16)$$

The expression $F(\delta_k, V^\star(k))$ being a random variable, equation (16) is indeed reminiscent of a random walk. The problem to be studied is how the random shocks δ_k and their probability distribution are transformed by the mapping F. To this end, we consider a given equilibrium interface $V^\star$ and inquire into the typical shape of the function $\delta \mapsto F(\delta, V^\star)$. Subsequently, it has also to be indicated how this function is affected by a change in the type of $V^\star$.

The initial equilibrium interface $V^\star$ underlying our numerical examples is similar to the interface $V' = \{V_o, V_1, V_2, V_3, V_4'\}$ in Figure 4. The important feature to note is that it connects horizontally to the hypotenuse of the limiting triangle Δ. Denote the corresponding equilibrium share of bold firms by b'. Then, consider a one-time negative shock to d, which instantaneously raises the payoff differential from its initial equilibrium value π' to π''. The resulting interface is $V'' = \{V_o, V_1, V_2, V_4''\}$ (before renumbering the vertices; the elimination of V_3 is here inessential). It has already been observed in the discussion of the loop-hysteresis at the end of Section 3 that a fall of π from π'' back to π' does not re-establish the previous population composition, i.e., π' and, thus, b' can no longer produce an equilibrium after the negative shock has materialized. In fact, the new equilibrium payoff must be smaller, and the new equilibrium share of bold firms must be larger (this is demonstrated in the Appendix).

Things are different if financial distress experiences a positive shock. Returning to the initial equilibrium interface $V' = \{V_o, V_1, V_2, V_3, V_4'\}$ in Figure 4, suppose the shock to d is limited in size, such that the resulting fall in the payoff does not wipe out vertex V_3. For example, let the shock give rise to $V = \{V_o, V_1, V_2, V_3, V_4\}$ as the new interface. In this case an increase in π back to the original level π' does set up the previous share b'

of bold firms, and the initial interface V' continues to constitute a state of equilibrium.

However, a new equilibrium comes into being if the positive shock to d is so strong that the fall in π eliminates a vertex (vertex V_3 in Figure 4). Similarly as before it can be shown that such an equilibrium is associated with a smaller share of bold firms, while its interface connects horizontally to the hypotenuse of Δ. At least for a certain range of shocks, the decrease in the equilibrium share brought about by a positive shock to d will be less than the increase resulting from a negative shock of the same size. This asymmetry in the equilibrium response is illustrated by the solid line in Figure 6. In terms of the discussion above, this curve is the graph of the function $\delta \mapsto b' + F(\delta, V')$ (b' und V' determining the original equilibrium). The zero slope for extreme values of δ is due to the fact that all shocks $\delta \geq 49\%$ induce the same degenerate interface, the instantaneous effect being a reduction of the payoff to such a low level that all firms are prudent. Likewise, all shocks $\delta \leq -29\%$ lead to an increase in π that temporarily induces all firms to turn bold.

The numerical data underying the solid line in Figure 6 correspond to Scenario 1 with its monotonic convergence. As a consequence, the new equilibrium share will be actually reached by the dynamic process after a one-time shock to d has occurred.

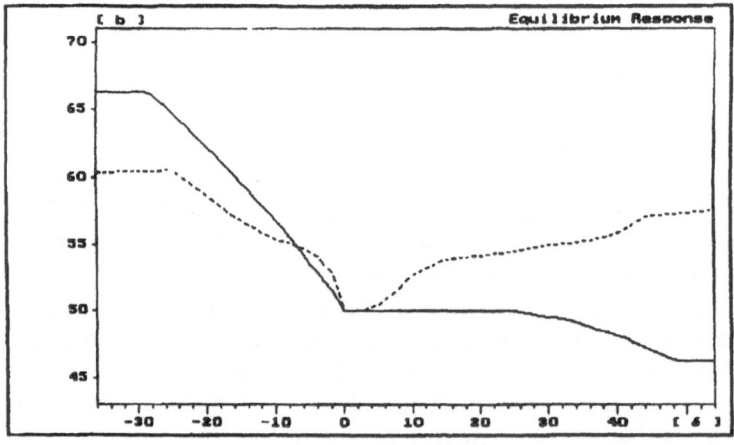

Figure 6 New equilibrium share (in %) of bold firms arising from a shock δ (in %) to financial distress; solid line represents Scenario 1, dashed line Scenario 2.

The case in which the original equilibrium interface has a vertical connection to the hypotenuse of Δ can be dealt with in an analogous way. The function $F(\delta, V')$ is again decreasing in δ, but now its constant segment extends over an interval of negative shocks, and the slope is steeper for positive δ. The qualitative reactions of the equilibrium share b^* of bold firms are summed up in Table 1.

Connection of original equil. interface to hypotenuse of Δ	Shock to financial distress d			
	weakly positive	strongly positive	weakly negative	strongly negative
horizontal	n.c.	$b^* \downarrow$ (H)	$b^* \uparrow$ (V)	$b^* \uparrow\uparrow$ (V)
vertical	$b^* \downarrow$ (H)	$b^* \downarrow\downarrow$ (H)	n.c.	$b^* \uparrow$ (V)

Table 1:   Response of the equilibrium share b^* of bold firms

Note: n.c. means no change, $\uparrow$ ($\uparrow\uparrow$) indicates a moderate (strong) reaction, letters H and V signify that the new equilibrium interface connects horizontally or vertically, respectively, to the hypotenuse of the limiting triangle Δ. Shocks to d are classified as strong (or weak) if they wipe out (or not) a vertex of the original equilibrium interface.

With this analysis we are sufficiently equipped to return to the stochastic process (16) describing the evolution of b^*. Since the equilibrium interfaces $V^*(k)$ change in the course of process (16), it is obvious that an assumption of identically and independently distributed shocks δ_k does not carry over to the innovations in b_k^*. The, also qualitatively, different effects of the δ_k, depending on whether the $V^*(k)$ are vertically or horizontally connected to the hypotenuse of Δ, do not even allow the transformed shocks $F(\delta_k, V^*(k))$ to be approximately i.i.d. Furthermore, their frequency distribution obtained from a sample simulation of (16) would show little resemblance to the frequency distribution of the causal shocks δ_k. This is, in particular, due to the possibility that bounded shocks δ_k leave the b_k^* unaffected. Hence the frequency of values around zero will be considerably higher for the induced shocks $F(\delta_k, V^*(k))$ than for the δ_k themselves.

The possibility of zero changes in b_k^* will also lead to more 'persistence' in the time series of system (16). A typical feature of a random walk for a variable x_k, say, are positive probabilities of successive perturbations in the same direction, so that over certain passages of time x_k is steadily increasing or decreasing. In process (16), the effect of successive shocks δ_k of the same sign is much more limited, or there may be no effect at all. For example, let the equilibrium interface $V^*(k)$ connect vertically to the hypotenuse of Δ and let a positive shock δ_k occur, which leads to a (moderate) fall in the share of bold firms, $b_{k+1}^* < b_k^*$. Since according to Table 1, $V^*(k+1)$ has a horizontal connection to the hypotenuse, a subsequent sequence of positive shocks, unless they wipe out a vertex, do not alter the equilibrium composition of the population of firms. The share of bold firms would be reduced if some of these shocks are stronger, but, as Figure 6 demonstrates, not proportionately so. A stronger reaction in b^* would only occur if the shocks changed their sign. Besides, this fact suggests that the innovations in b_k^* are weakly negatively autocorrelated.

To illustrate these remarks, Scenario 1 of process $(3-10)$ has been simulated under the assumption that a shock to financial distress occurs every 4 years.[13] This time span was long enough for the system to converge close to the, old or new, equilibrium position. A sample of the resulting time series of the aggregate capital growth rate g^k is shown in the upper panel of Figure 7. By virtue of the delayed adjustments of g^k in equation (5), this series is smoother than the series of the actual population share b_t or the stepwise movements of the equilibrium share b_t^*. Three 'growth regimes' can be clearly distinguished, though the numerical differences in the growth rates are not too large. In the interval between (roughly) $t=15$ and $t=50$, the growth rate g^k hovers around a trend value of about 3.8% (in the initial equilibrium, $g^k = 3.5\%$ prevailed), in the time interval $[105, 130]$ trend growth is slightly above 3.6%, and in the interval $[140, 160]$ it is slightly above 3.9%.

It is interesting to note that the pure random walk that is generated by this sample of shocks: $x_{4(k+1)} = x_{4k} - \delta_{4k}$, exhibits a pattern which is rather distinct from the capital growth rates.[14] In the sample here presented, the x-series happens to perform a spurious oscillatory motion; from $t=16$ until $t = 136$ (i.e., $k = 4$ until $k = 34$), it is an only mildly perturbed trough-to-trough movement, which at $t = 76$ and $t = 92$ attains two local peak values of nearly the same height (they are also visible in the g^k-series in Figure 7, but there the two peaks appreciably differ). As a consequence, the sequence of the shocks δ_{4k} with its 42 members displays a positive first-order serial correlation of $\rho_1 = 0.219$. By contrast, the corresponding sequence of innovations in the equilibrium share b^* of bold firms has a negative serial correlation, $\rho_1 = -0.202$, where 25 times the shocks δ_{4k}

---

[13]Each shock independently drawn from a normal distribution with a standard deviation of 10%.

[14]The choice of the indices of x makes the time scale of process $(3-10)$, equation (16), and the original random walk comparable; δ_{4k} is the shock to d at time $t=4k$ ($k = 0, 1, 2, \ldots$) and the minus sign is employed to let x and b^* move in the same direction.

have no impact at all. These observations may here suffice to point out that a pure random walk has a significantly different character from the hysteresis in the dynamics (3–10), when, being universally stable with monotonic convergence, this process is subjected to repeated infrequent shocks.

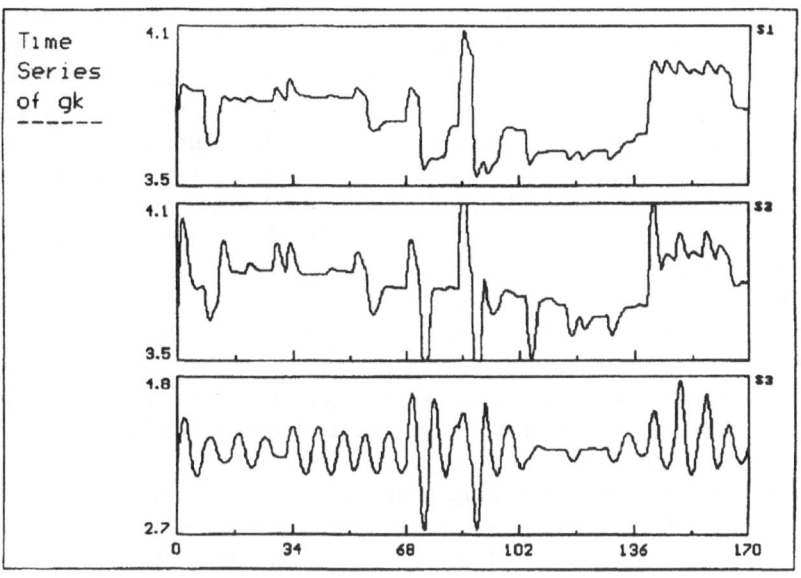

Figure 7 Time series of the aggregate growth rate g^k (in %) of fixed capital, resulting from the same sequence of repeated random shocks in these scenarios $S1, S2$ and $S3$

6 Hysteresis in Cyclical Scenarios

In this section it will first be seen that even if the assumption of universal stability is maintained, the consequences of a perturbation of the steady state may be markedly different if convergence is cyclical. We may again invoke Figure 4 and let a starting equilibrium interface have the shape of $V' = \{V_o, V_1, V_2, V_3, V_4'\}$ with its horizontal connection to the hypotenuse of

Δ. Consider then a positive shock to d which drives the payoff from π' down to π_o. The shock-induced interface being given by $V = \{V_o, V_1, V_2, V_3, V_4\}$, it has already been noted that this does not change the equilibrium position. If direct convergence prevails, the process would therefore end up in the initial interface V'. Under cyclical stability, however, the upward motion of the payoff will not come to a halt at π' but will overshoot this equilibrium value.[15]

The point is that once the payoff has passed π', the situation resembles that of a negative shock occurring to the original steady state, i.e., from then on V' no longer constitutes an equilibrium. Denoting the interface arising from the payoff π_t at time t by $V(t)$, we can utilize (essentially) the same argument as in the previous section to infer an increment in the corresponding equilibrium share of bold firms: $\pi_t > \pi'$ implies $b^*(V(t)) > b^*(V')$. Moreover, $b^*(V(t))$ increases as long as π_t continues to rise. When π_t begins to fall, for example after a peak at $\pi = \pi''$ in Figure 4 with the interface $V'' = \{V_o, V_1, V_2, V_4''\}$, the equilibrium share stays constant: $b^*(V(t)) = b^*(V'')$. Some of the previous increase in b^* is undone after the next overshooting, when π_t passes the payoff level corresponding to $b^*(V'')$. The equilibrium share starts to rise again after the payoff has reached a lower turning point, *etc.*, until these oscillations will have died out. In contrast to direct convergence, where a positive shock to financial distress has a zero or negative impact on the final level of b^*, under cyclical convergence this effect is positive.

As mentioned, this reasoning includes the case of a negative shock to distress, so that it, too, raises the equilibrium share of bold firms in the new state of rest. Varying the strength of the shocks to d in the numerical example of our Scenario 2 and computing the resulting new equilibria yields the second response function of b^* in Figure 6, which is depicted as the dashed line. The effects in the cyclical scenario depend heavily on the first peak and trough values after the shock. Hence the fact that the reactions to the negative shocks are not unambigously stronger or weaker than in Scenario 1 must be ascribed to the interplay of the variables in the dynamic process as a whole. Some explanations could be given for the specific case under consideration, but they seem hard to generalize.

A mirror-image argument to the process above establishes that if the initial equilibrium interface is vertically connected to the hypotenuse of Δ, then all shocks to d, again positive and negative shocks alike, reduce the share of bold firms in the new equilibrium. While Figure 6 suggests that b^* is a strictly increasing (or decreasing, respectively) function of the size of the shocks δ, there are also examples to be found where this relationship is

---

[15]As may be expected from the weak amplifying forces under cyclical stability, the downward jump in the payoff caused by the initial positive shock actually initiates a further decrease in π for a short while, before the stabilizing mechanism sets in.

not monotonic; starting from an initial equilibrium with vertical connection to the hypotenuse, b^* may first decrease as δ rises from zero, and then increase again as the shocks become stronger. We nevertheless do not wish to overemphasize this possibility. It is remarkable enough that the steady state growth rate will react in the same direction, whether there were a positive or negative shock to financial distress, and that the direction itself depends on the history of the fluctuations of the economy, i.e., whether the previous turning points have generated a vertical or horizontal connection to the hypotenuse of Δ.[16]

After studying the equilibrium response, the system may be subjected to the same sequence of financial shocks as in Scenario 1. Generally the oscillations in Scenario 2 make convergence to an equilibrium slower than under monotonic convergence. Since within the assumed four years between two successive shocks the adjustment process still bridges most of the initial gap between the actual and the equilibrium share of bold firms, this timing of the shocks was maintained. The resulting time series of the capital growth rate g^k is shown in the middle panel of Figure 7. The three growth regimes of the first scenario are still discernible. On the other hand, the shocks produce some amplifying effects, though they are only temporary (note that the scale is the same for the two panels of Scenario 1 and 2). The respective average levels of the growth rates in the three regimes are barely affected, either. In addition, the sequence of the innovations $F(\delta_k, V^*(k))$ to the b_k^* in process (16) exhibits nearly the same order of negative serial correlation, namely, $\rho_1 = -0.219$. To summarize, although the equilibrium response functions are fairly dissimilar in Scenario 1 and 2, these differences tend to be washed out in the time series of the variables generated by the repeated random shocks. The main reason are, of course, the recurrent changes in the type of the equilibrium interfaces, from a vertical to a horizontal connection and *vice versa*.

A difference between the two stochastic processes of Scenario 1 and 2 is obtained in the frequency distribution of the innovations $F(\delta_k, V^*(k))$. Comparing the two equilibrium reactions in Figure 6, it is immediate that in Scenario 2 zero changes, or nearly-zero changes for that matter, will be observed less often. This implies a greater spread in the frequency distribution around zero and, in a histogram, makes it look a bit more like the distribution of the causal shocks δ_k.

Another property of the frequency distributions arising from the two scenarios is associated with the slope of the equilibrium response functions.

---

[16]If an equilibrium position with payoff π^* has been approached by way of dampened oscillations, there will be many vertices close together in a vicinity of $(\lambda, \mu) = (\pi^*, \pi^*)$ (or infinitely many in a mathematical abstraction). Whether such an interface is classified as having a vertical or horizontal connection to the hypotenuse of Δ will depend on the shape that remains when the 'nearest' vertices to (π^*, π^*) are wiped out by the shock.

Figure 6 illustrates that the shocks in the cyclical scenario, in one direction at least, have a weaker impact on the equilibrium share b^* than in Scenario 1. This finds its expression in the fact that the simulations of Scenario 2 yield a distribution of the $F(\delta_k, V^*(k))$ with a lower standard deviation (the exact values are 1.29% in Scenario 2 *versus* 2.27% in Scenario 1).

We may now address the problem of instability and ask what happens if the process does not converge to an equilibrium. It is easily seen that system (3–10) has built in a global negative feedback mechanism, which prevents the trajectories from exploding. If, for example, all firms adopt the set of bold strategies for a longer while, then sooner or later the corresponding aggregate exposure (equation (8)) dampens the psychology of financial markets and raises financial distress d to similarly excessive levels (see (9); at least if, as here assumed, the growth rate coefficient α_d in the distress function (14) is not too large). This rise in d inexorably reduces the differential payoffs from bold behaviour in (10), and those firms with the highest λ-switching values return to the prudent strategies. An analogous argument applies should all firms have become prudent. If, on the other hand, no equilibrium is approached, it follows that the economy must undergo persistent fluctuations. From a Minskian point of view, which stresses the influence of financial markets on real activity, these oscillations might even be seen as a constituent part of a theory of endogenous business cycles.

It has been noted before that when a vertex is wiped out, the process gives rise to a discontinuity in the increments of the share of bold firms from one short period to another. One may surmise that this switching in the regimes of the ordinary difference equations leads to chaotic dynamics. It turned out, however, that the sudden changes were not so serious in this respect. In all our simulation runs the motions soon developed into regular oscillations which, for all practical purposes, can be described as strictly periodic. Thus, the economy no more converges to a point of equilibrium, but to a closed cycle (or a closed orbit, more technically). Similarly as with the shifting equilibria above, it will be expected that these cycles are not unique and that different shocks induce convergence to different cycles.

A condition for the equilibria of the process to become unstable is that the growth effects in the two functions f_π and f_d in equations (14) and (15), as measured by the slope coefficients α_π and α_d of the perceived rate of growth, are sufficiently strong. An example is Scenario 3, which was made explicit in the preceding section. Let us then perform the same experiment as with the previous two scenarios and perturb the steady state position by a shock to financial distress. The first remarkable effect to be observed is the phenomenon of corridor stability (as Leijonhufvud has called it). A positive shock does not change the equilibrium, which is already known. It turns out that its stability is preserved when the shock is small enough, where convergence is virtually monotonic. The critical value of the perturbation is about $\delta = 0.98\%$. If δ exceeds this benchmark, then in its adjustment back to π^* the payoff overshoots this initial equilibrium value. Unlike Scenario 2, the acceleration in the oscillations then setting in is so vigorous that they are no longer dampened. As indicated, these fluctuations are attracted by a periodic orbit. Interestingly, negative shocks, which do affect the equilibrium position, are totally destabilizing, i.e., the periodic cycles come into being however small in modulus the shocks may be. The corridor stability arising in this way is therefore one-sided, as we may say.

Under sustained growth cycles, a unique steady state position in ordinary dynamic systems, though not manifest in the time series, has still a role to play as a state of reference. Since furthermore the equilibrium values are often a satisfactory proxy for the time averages of the variables, the steady state provides a theoretical and, to some extent, even practical basis of the analysis. In the present dynamics, the steady state which we used for reference is, and remains, not only virtual, it is also continuously shifting. Such a system has no more firm anchor. A notion of equilibrium is maintained, but it is devoid of any deeper meaning.

Besides the time averages of selected variables, another characterization of the limit cycles are the peak and trough values of the share of bold firms. Upon systematic variations of the shocks δ, both of them are seen to increase weakly as the (positive and negative) shocks increase in size; the troughs somewhat more than the peaks, so that the amplitude slightly decreases. This is shown by the upper and lower lines in Figure 8. Additional details of the outcome of the present calibration are given in the Appendix.

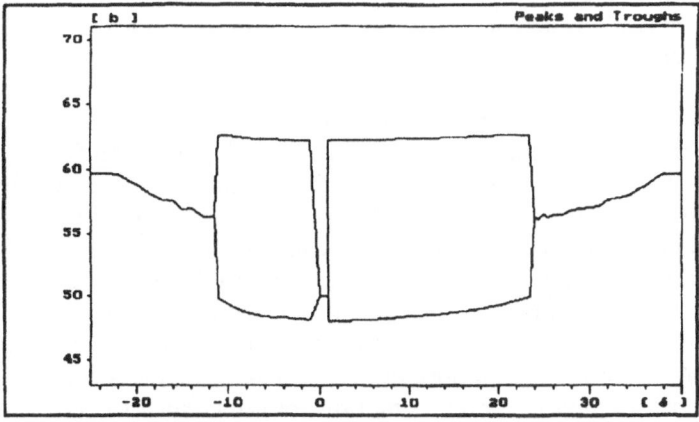

Figure 8 Share of bold firms (in %): peak and though values of the limit cycles in Scenario 3 induced by shock δ (in %); convergence to a new steady state if they coincide.

A striking phenomenon emerges when the shocks grow large in the experiment, that is, if $\delta \geq 24\%$ for positive shocks and $\delta \leq -11.5\%$ for negative shocks. After a shock in this range, the variables still oscillate in the expected manner for quite some time, even six or seven cycles in some cases, with a slowly diminishing amplitude. From the experience with the weaker shocks, one is inclined to think that again the economy is going to approach a periodic motion. Then, however, although no further vertex is eliminated, the amplitude is more drastically reduced — and the economy converges to a single state of equilibrium! The system thus exhibits an extended kind of corridor stability.

To venture an analogy, it seems as if the economy needs a long run-up to gain such a momentum and enter so deep into the force field of a planet that, after a number of narrowing orbits, it is able to escape from the other forces of attraction in space, so that the planet's forces of gravity finally succeed. In particular, convergence to an equilibrium is achieved by extreme shocks which obliterate history and wipe out all previous vertices, even temporarily causing all firms to adopt the same type of behaviour (though these features are not a necessary condition for the phenomenon to occur). The analogy with the gravitation field is also valid insofar as the larger shocks speed up the process of convergence. As far as the author knows, such a reswitching of (point-) stability, brought about by sufficiently *strong* shocks, has not been obtained before in economic dynamic systems.

The different shock effects produced by the variations of δ are summarized in Figure 8. Clearly, the intervals where the upper line for the peak values and the lower line for the trough values of b coincide, signify that here the economy converges to a single steady state position. This happens in the small corridor between $\delta = 0$ and $\delta = 0.98\%$, and for the sizeable shocks just made explicit. Reading the diagram from the middle to the outer range of δ, we see the reswitching of steady state stability.

The economy of Scenario 3 was, likewise, subjected to the above financial random shocks. Maintaining the timing of their arrival every four years means that the oscillations are perturbed roughly every half cycle, irrespective of the particular stage in which the economy may presently be (the fluctuations have a period between 7 and 8 years). Even if a sizeable shock to d were to produce convergence to a single equilibrium, the adjustments are cyclical and take some time. So it is quite likely that one of the subsequent shocks neutralizes the gravitation towards the equilibrium point, and the attraction of a closed orbit again becomes the dominant force. Considering the latter as the normal case, the single shocks can be viewed as giving rise to another periodic motion towards which the economy begins to converge. On the analogy of the shifting steady states, hysteresis in this framework means that these periodic cycles shift over time. For short, one may speak of *cyclical hysteresis*.

It is, however, to be expected that this kind of hysteresis is more difficult to perceive. One reason is that the cycles do not appear to differ very much in their characteristics. The variations in the peak and trough values of b in Figure 8 are relatively small as the shocks δ are systematically increased (the scale of b in Figures 6 and 8 is the same). On the other hand, the shocks in the dynamic economy occur rather frequently in comparison to the time required to approach the limit cycle. These two features combined may cause the noise of the shocks to dominate the changes in the true cycle characteristics, to which the system would converge if it had been given enough time. Or, if we study the moving time averages of a smooth cyclical series like g^k, there will be only a limited scope for them to evolve as being governed by a random walk.

This supposition is essentially brought out by the shock-generated time series of the aggregate capital growth rate g^k in the bottom panel of Figure 7. Contrasting it with the other two scenarios, the larger scale in the third panel should be noticed, which is due to the wider fluctuations in this scenario. Especially the shocks between $t = 70$ and $t = 90$ have a much stronger effect on the amplitude. The different pattern of the bottom series makes it also more difficult to recognize the three growth regimes of the previous scenarios. Referring to the time averages of g^k, however, two of them are still present with approximately the same growth rates as before, though they stand out less clearly as in the top panel. These are the first and third regime, i.e., the time intervals $[15, 50]$ and $[140, 160]$. On the other hand, the growth regime over the period $[105, 130]$ with its distinctly lower growth rate in the upper two panels has disappeared; here average growth takes place at almost the same level as in the first interval.

Moreover, quite surprisingly when looking at the first hundred 'years', and knowing that these cycles are basically endogenous, temporarily the oscillations nearly vanish. This phenomenon can be explained on the basis of the discussion of Figure 8. A strong positive shock to financial distress has precipitated a deep fall in g^k around $t = 90$, such that, if left to itself, the system would converge to a single equilibrium position. It happens that the subsequent shocks do not decisively disturb these adjustments. A steady state growth path is actually reached 20 time units later. Then, around $t = 115$, another positive shock to d arrives. Here the Leijonhufvud corridor stability criterion applies. The shock is so small that the eonomy returns to a new steady state, with an only slightly higher rate of growth. From $t = 128$ on, the shocks are in the 'normal' range and the conomy resumes its cyclical behaviour.

Apart from hysteresis in the — only mildly — shifting time averages of g^k, it might be said that the process also displays *hysteresis in kind* : from sustained oscillations to steady growth, and back to sustained cycles again. Though, this switching in qualitative behaviour will not be expected to take place too often.

7 Conclusion

Adopting the framework which was discovered by Cross (1994) in the physical sciences and translated into economic terms, the present paper has put forward a model of heterogeneous investors in fixed capital with discontinuous, lumpy adjustments to the changes in the real sector. It was demonstrated how these individual inertia carry over to the macro level and lead to path-dependence in the aggregate variables. The dynamic features of the model are richer than what may be obtained by direct hypotheses about the macroeconomic relationships, or by employing the representative agent. Our approach thus widens the scope for the notion of hysteresis. In particular, the often used random walk processes can here only serve as an approximation to the model (within a certain setting), and possibly not a good one.

The individual behaviour of agents is based on a number of simplifying assumptions. It will, however, not be expected that the hysteresis effects observed would be weakened if they were relaxed. It may rather be worthwhile to look for more differentiated or even new effects. Three extensions of the model present themselves for future research in this direction. (a) The uniform distribution of the agents' characteristics may be replaced with other, less regular distributions. This could reinforce asymmetric reaction patterns. (b) The individual (λ, μ)-switching values which have been treated as constants may be endogenized, so that their distribution across agents is shifting over time. (c) Also more than two polar types of agents' strategies may be examined.

We wish to point out that our analysis may have a bearing on macrodynamic modelling. Since not every medium- or long-run model can afford to consider the micro level in similar detail as we did, it may be asked if the error committed by disregarding the corresponding hysteresis effects is negligible. According to the discussion in the last two sections an answer will not be independent of the kind of dynamics the model builder is interested in. Put briefly, we have found that convergence to the steady state positions strongly favours persistence in the variables in response to exogenous shocks, whereas endogenously generated fluctuations tend to absorb these shocks to the system. A tentative conclusion is that macro models whose deterministic part gives rise to sustained (nonchaotic) cyclical behaviour are more robust to hysteresis originating at the micro level than models with stable equilibria.

8 Appendix

Given $V = \{V_o, V_1, \ldots, V_m\} = \{(\lambda_o, \mu_o), (\lambda_1, \mu_1), \ldots, (\lambda_m, \mu_m)\}$ and a (new) value of the payoff, π, distinguish two cases in determining the corresponding new interface.

Case 1, $\pi \leq \lambda_m$: Determine $k = \min\{i : \pi \leq \lambda_i\}$ and set

$$\Phi(\pi, V) = \{(\lambda_o, \mu_o), (\lambda_1, \mu_1), \ldots, (\lambda_{k-1}, \mu_{k-1}), (\max[\pi, \lambda_{min}], \mu_k)\}$$

As for $\pi \leq \lambda_{min}$, recall $\lambda_o = \lambda_{min}$, so that $\Phi(\pi, V) = \{(\lambda_o, \mu_o)\} = \{(\lambda_{min}, \mu_{max})\}$ for these extreme payoffs.

Case 2, $\pi > \lambda_m$: Determine

$$k = \begin{cases} \min\{i : \mu_i \leq \pi\} & \text{if } \mu_m \leq \pi \\ m+1 & \text{else} \end{cases}$$

and set

$$\begin{aligned}
\Phi(\pi, V) &= \{(\lambda_o, \mu_o), (\lambda_1, \mu_1), \ldots, (\lambda_{k-1}, \mu_{k-1}), \\
&\quad (\min[\pi, \mu_{max}], \min[\pi, \mu_{max}])\}
\end{aligned}$$

In particular, for $\pi \geq \mu_{max}$ one has $\Phi(\pi, V) = \{(\lambda_{min}, \mu_{max}), (\mu_{max}, \mu_{max})\}$.

Derivation of the Investment Growth Rate g_t^i in Equation (6):

Given the length of the adjustment period h, let hI_t be the volume of aggregate net investment over the time interval $[t, t+h)$ and K_t the capital stock at time t. Specify the (annualized) growth rates $g_t^k = (K_{t+h} - K_t)/hK_t$ and $g_t^i = (I_t - I_{t-h})/hI_{t-h}$. Using $I_t = (K_{t+h} - K_t)/h = g_t^k K_t$, one computes

$$\begin{aligned}
g_t^i &= \frac{g_t^k K_t - g_{t-h}^k K_{t-h}}{h\, g_{t-h}^k K_{t-h}} = \frac{(g_t^k - g_{t-h}^k)K_t + g_{t-h}^k(K_t - K_{t-h})}{h\, g_{t-h}^k K_{t-h}} \\
&= \frac{g_t^k - g_{t-h}^k}{h\, g_{t-h}^k}\, \frac{K_t}{K_{t-h}} + \frac{K_t - K_{t-h}}{h\, K_{t-h}} \\
&= \frac{g_t^k - g_{t-h}^k}{h\, g_{t-h}^k}\, (1 + h\frac{K_t - K_{t-h}}{hK_{t-h}}) + g_{t-h}^k
\end{aligned}$$

which is equation (6).

Specification of the Output Growth Rate $f_y = f_y(g^i)$ in Equation (7):

With respect to a given value of $\bar{g}$ and the four parameters $\gamma_{y1}, \gamma_{y2}, \alpha_{y1}, \alpha_{y2}$, where $0 \leq \gamma_{y1} < \gamma_{y2}$, $0 \leq \alpha_{y2} < \alpha_{y1} \leq 1$, define

$$f_y(g^i) = \begin{cases}
g^i & \text{if } |g^i - \bar{g}| \leq \gamma_{y1} \\
\bar{g} + \gamma_{y1} + \alpha_{y1}(g^i - \bar{g} - \gamma_{y1}) & \text{if } \gamma_{y1} \leq g^i - \bar{g} \leq \gamma_{y2} \\
\bar{g} - \gamma_{y1} + \alpha_{y1}(g^i - \bar{g} + \gamma_{y1}) & \text{if } -\gamma_{y2} \leq g^i - \bar{g} \leq -\gamma_{y1} \\
\bar{g} + \gamma_{y1} + \alpha_{y1}(\gamma_{y2} - \gamma_{y1}) \\
\quad + \alpha_{y2}(g^i - \bar{g} - \gamma_{y2}) & \text{if } g^i \geq \bar{g} + \gamma_{y2} \\
\bar{g} - \gamma_{y1} - \alpha_{y1}(\gamma_{y2} - \gamma_{y1}) \\
\quad + \alpha_{y2}(g^i - \bar{g} + \gamma_{y2}) & \text{if } g^i \leq \bar{g} - \gamma_{y2}
\end{cases}$$

Thus, $f_y(g^i)$ has slope 1 over the interval $[\bar{g}-\gamma_{y1}, \bar{g}+\gamma_{y1}]$, slope α_{y1} over the two intervals $[\bar{g}-\gamma_{y2}, \bar{g}-\gamma_{y1}]$ and $[\bar{g}+\gamma_{y1}, \bar{g}+\gamma_{y2}]$, and slope α_{y2} else.

Numerical Details of the Simulations:

$\bar{g}$	=	0.03	γ_{y1} = 0.03		γ_{y2} = 0.06	
β_g	=	2.00	α_{y1} = 0.667		α_{y2} = 0.25	
β_k	=	1.00	g_p = 0.01		g_b = 0.06	
β_d	=	1.00			h = 0.05	

On the basis of $\lambda_{min} = 0$ and $\mu_{max} = 10$, the initial equilibrium interface V^*, which accomplishes $b^* = B(V^*) = 50\%$, is given by

$$V^* = \{(0,10), (1,9), (2,8), (3,7), (\pi^*, \pi^*)\}, \qquad \pi^* = 6.3166$$

The special value of π^* (the same in the three scenarios) is brought about by the parameters $\sigma_\pi = 10$, $\psi = 0.30$, $\gamma_d = \alpha_d \bar{g}$ in the payoff and distress functions (14, 15), and

α_π	=	0	α_d	=	0	γ_π	=	2.81662
α_π	=	4	α_d	=	1	γ_π	=	2.11662
α_π	=	5	α_d	=	2.5	γ_π	=	1.94162

in Scenario 1, 2 and 3, respectively.

Discussion of Equation (16) in Scenario 1

With respect to the discussion following equation (16), it has to be shown that in response to a negative shock to d, the payoff in the new equilibrium is smaller, and the new equilibrium share of bold firms is larger. To see this, refer to the value $\pi^* = \pi^*(b)$ in equation (11), which the payoff must attain in equilibrium. In the initial equilibrium we have $\pi' = \pi^*(b')$, $V' = \Phi(\pi', V')$, and the fixed-point equation (12) reads $b' = B[\Phi(\pi', V')] = B[\Phi(\pi^*(b'), V')]$. With respect to the shock-induced interface V'', the strict inequality $B[\Phi(\pi, V'')] > b' = B[\Phi(\pi', V')]$ holds for all payoffs $\pi \geq \pi'$.[17] In particular, $B[\Phi(\pi^*(b'), V'')] > B[\Phi(\pi^*(b'), V')] = b'$. Since $b \mapsto B[\Phi(\pi^*(b), V'')]$ is a decreasing function (cf. the remark following equation (12)), the solution $\tilde{b}$ of the new fixed-point equation $B[\Phi(\pi^*(b), V'')] = b$ must exceed the original share b'. Since we have $\pi^*(\tilde{b}) < \pi' < \pi''$, i.e., the payoff has to fall from π'' to its new equilibrium level, it follows that the equilibrium interface changes its shape and is now vertically linked to the hypotenuse of the Δ-triangle.

Closed Orbits in Scenario 3:

Besides the peaks and troughs of the share of bold firms mentioned in the text, the present calibration of the model in Scenario 3 yields the following features of a typical cycle. The trough values of the perceived rate of growth g, the aggregate capital growth rate g^k, and the growth rate of investment g^i (a second derivative so to speak) are, respectively, -1.42%, 3.46%, and -3.68% per unit of time; the corresponding peaks are 8.67%, 4.04%, and 11.95% (recall that practically all growth rates are instantaneous). The

---

[17]For example, the interface in Figure 4 that now corresponds to π' is given by $\tilde{V} = \{V_o, V_1, V_2, \tilde{V}_4\}$. Clearly, the area below $\tilde{V}$ is larger than the area below $V' = \{V_o, V_1, V_2, V_3, V_4'\}$, which equals $B[\Phi(\pi', V')]$ (apart from the normalization).

period of the cycle is 7.55 time units, where the phases from peak to trough and from trough to peak are almost symmetric. The fluctuations in g and g^i are not too unrealistic compared to ordinary business cycles if the underlying time unit is taken as a year. A shortcoming in this business cycle interpretation is the small amplitude in the capital growth rate — unless the concept of g^k is refined and g^k is viewed as an aggregate *desired* growth rate of fixed capital. Since already the mean $\tilde{g} := g_b b + (1-b)g_p$ was an aggregate target rate of growth, $\tilde{g}$ and g^k would be associated with two different time horizons. So it has to be admitted that this interpretative construction is somewhat artificial. If beyond the purely theoretical thrust of the model, calibration of these cycles is seen as an important task to validate the model, then a more detailed specification of the adjustments in the real sector is demanded.

References

[1] Blanchard, O.J. and Summers, L.H. (1986), "Hysteresis and the European unemployment problem", in S. Fischer (ed.), *NBER Macroeconomics Annual 1986*. Cambridge: MIT Press; pp. 15–78

[2] Blanchard, O.J. and Summers, L.H. (1987), "Hysteresis in unemployment", *European Economic Review*, 31, 288–295

[3] Cross, R. (1994), "The macroeconomic consequences of discontinuous adjustments: selective memory of non-dominated extrema", *Scottish Journal of Political Economy*, 41, 212–221

[4] Dixit, A. (1992), "Investment and hysteresis", *Journal of Economic Perspectives*, 6, 107–132

[5] Dixit, A. (1995), "Irreversible investment with uncertainty and scale economics", *Journal of Economic Dynamics and Control*, 19, 327–350

[6] Franke, R. (1987), *Production Prices and Dynamical Processes of the Gravitation of Market Prices*. Frankfurt a.M.: P. Lang

[7] Franke, R. (1994), "A reappraisal of adaptive expectations", *mimeo*, University of Bremen

[8] Franz, W. (1990), "Hysteresis in economic relationships: an overview", *Empirical Economics*, 15, 109–125

[9] Giavazzi, F. and Wyplosz, C. (1985), "The zero root problem: a note on the dynamic determination of the stationary equilibrium in linear models", *Review of Economic Studies*, 52, 353–357

[10] HEINER, R.A. (1988), "The necessity of delaying economic adjustment", *Journal of Economic Behavior and Organization*, 10, 255–286

[11] Mayergoyz, I.D. (1991), *Mathematical Models of Hysteresis*. New York: Springer

[12] Miller, M.H. and Orr, D. (1966), "A model of the demand for money by firms", *Quarterly Journal of Economics*, 80, 413–435

Non Linear Dynamics and Utility Functions in Overlapping Generations Models

Gilles DUFRÉNOT and Laurent MATHIEU

1   Introduction

Grandmont [7] was among the first to study the possibility of self-sustaining cycles in the overlapping generations models. In his paper, the origin of these cycles is the conflict between the substitution effect and the income effect due to variations of relative prices. Grandmont's approach has been criticized on empirical basis : income effects in the model are too large compared to their values in real life. Our paper, conversely, provides some theoretical arguments that explains why cyclical and complex paths cannot be ruled out. In this view, we examine the problem of endogenous fluctuations in pure exchange economies from the viewpoint of macroeconomic theories of consumption and saving.

Many studies explore the idea that in the face of uncertainty concerning anticipated labor income, consumers can be more or less prudent (Carrol and Summers [2], Carrol [1]). Also, the role of risk aversion has been investigated (Shapiro [9], Dynan [4]). The propensity to consume out of labor income depends on both factors which appear as potential sources of fluctuations in aggregate consumption and saving. Moreover, recent works exploring the importance of these factors have reached the conclusion that the combination of risk aversion and prudent behavior may be a major cause of the very unstable nature of aggregate saving (Dubois and Bonnet [3]). These observations motivate our analysis.

Following an early suggestion by Dreze and Modigliani [5], we show that the degree of curvature of an offer curve (or equivalently, the substitution and income effects) can be expressed in terms of the first, second, and third derivatives of the utility functions that describes preferences of the consumers. Those derivatives are considered for they contain information on prudent behaviors and attitudes towards risk in the face of uncertainty. To achieve this result, we start from the methodology initiated by Sato

[8] which shows a close connection between the hypotheses made on the Slutsky equations corresponding to excess demand functions and classes of utility functions from which cycles and complex dynamics emerge.

The paper is organized as follows. Section 2 presents the model. In Section 3 we derive classes of utility functions. Section 4 uses some simulations to illustrate the nonlinear dynamics of consumption associated to each type of utility functions.

2   The Model

This model, which is based on an overlapping generations model, is composed of two distinct agents each living two periods. In the first period the agent is in the younger generation; while in the second he is in the older generation. In this pure exchange economy, each agent receives units of a good which is neither produced nor stored. The population is constant and both generations engage in goods exchanges. The market driven by a system of relative prices. The model is based on the following assumptions. There is a representative agent in each generation who determines how much to consume in the current period c_t^y and in the future period c_{t+1}^o. The representatives' utility function $U\left(c_t^y, c_{t+1}^o\right)$ is additively separable and satisfies the usual assumptions (monotonicity and concavity). The endowment in the two periods of his life are given as $\omega = (\omega^y, \omega^o)$.

Based on these assumptions, the structure of the economy is represented by the following two equations.

- Following Gale [6], a program is called feasible if the aggregate consumption equals aggregate endowments in each period:

$$c_t^y + c_t^v = \omega^y + \omega^o \tag{1}$$

- A program is called competitive if a young agent maximises his utility under the budget constraint:

$$\begin{cases} Max \ U\left(c_t^y, c_{t+1}^o\right) = U_1\left(c_t^y\right) + U_2\left(c_{t+1}^o\right) \\ p_t c_t^y + p_{t+1} c_{t+1}^o = p_t \omega^y + p_{t+1} \omega^o \end{cases} \tag{2}$$

The price of the good in period t is $(p_t)_{t=1,2}$. The utility functions of the young and the old agents are respectively $U_1\left(c_t^y\right)$ and $U_2\left(c_{t+1}^o\right)$. These functions satisfy the following conditions:

$$U_i'\left(.\right) > 0 \quad \text{and} \quad U_i''\left(.\right) < 0 \quad \text{for} \quad i = 1, 2. \tag{3}$$

Let us define the interest factor ρ_t as the rate of exchange between current and future consumption:

$$\rho_t = \frac{p_t}{p_{t+1}} = 1 + i_t \tag{4}$$

where i_t is the real interest rate. In order to simplify the resolution of the model, we define $\mu_t = \frac{1}{\rho_t}$. Thus, the system (2) is written:

$$\left\{ \begin{array}{l} Max \ U\left(c_t^y, c_{t+1}^o\right) = U_1\left(c_t^y\right) + U_2\left(c_{t+1}^o\right) \\ \left(\omega^y - c_t^y\right) + \mu_t\left(\omega^o - c_{t+1}^o\right) = 0 \end{array} \right. \tag{5}$$

- We are interested by the consumption time paths which satisfy both equations (1) and (5). The first order optimality conditions of the maximisation program implies:

$$\mu_t = \frac{U_2'}{U_1'} \tag{6}$$

where U_1' and U_2' are the first order partial derivatives of U relatively to c_t^y and c_{t+1}^o. Substituting this last equation into the budget constraint implies:

$$\frac{U_2'}{U_1'} = \frac{\omega^o - c_{t+1}^o}{c_t^y - \omega^y} \tag{7}$$

Our approach to the study of the dynamics of aggregate consumption, given the specifications of the utility functions, is based on two steps. Firstly, we will use the methodology proposed by Sato [8] which consists of deriving a class of additive utility functions in accordance with the assumptions of the substitution and income effects of the Slutsky equations. Secondly, we will use the utility function to study the behavior of the consumption dynamics.

3 Derivation of the Utility Functions

We first write the Slutsky equations, then in deriving the utility functions we impose different restrictions on the terms representing the substitution and income effects: either the price elasticity of the hicksian demand is constant or the engel curves are linear.

3.1 THE SLUTSKY EQUATIONS

The Lagrangian of the consumer program (5) is:

$$L\left(c_t^y, c_{t+1}^o, \lambda\right) = U_1\left(c_t^y\right) + U_2\left(c_{t+1}^o\right) + \lambda\left[\omega^y - c_t^y + \mu_t\left(\omega^o - c_{t+1}^o\right)\right] \tag{8}$$

where λ is the Lagrange multiplier. The first-order conditions are:

$$\left\{ \begin{array}{l} U_1'\left(c_t^y\right) - \lambda = 0 \\ U_2'\left(c_{t+1}^o\right) - \lambda\mu_t = 0 \\ \omega^y - c_t^y + \mu_t\omega^o - \mu_t c_{t+1}^o = 0 \end{array} \right. \tag{9}$$

The solution of this system is a maximum if the determinant of the bordered hessian matrix HB is positive. This matrix is written as follows:

$$HB = \begin{pmatrix} U_1'' & 0 & -1 \\ 0 & U_2'' & -\mu_t \\ -1 & -\mu_t & 0 \end{pmatrix} \tag{10}$$

and the determinant is:

$$|HB| = -\mu_t^2 U_1'' - U_2'' > 0 \tag{11}$$

Respectively, U_1'' and U_2'' are the second order derivatives of $U_1(c_t^y)$ and $U_2(c_{t+1}^o)$ relative to c_t^y and c_{t+1}^o.

The determinant is positive since we have assumed that the utility function is strictly concave. The differentiation of the system (9) leads to:

$$\begin{cases} U_1'' dc_t^y - d\lambda = 0 \\ U_2'' dc_{t+1}^o - \lambda d\mu_t - \mu_t d\lambda = 0 \\ d\omega^y - dc_t^y + \mu_t d\omega^o + \omega^o d\mu_t - \mu_t dc_{t+1}^o - c_{t+1}^o d\mu_t = 0 \end{cases} \tag{12}$$

from which we deduce the expression of the Slutsky equations (see appendix 1 for the proof):

$$\begin{cases} \frac{dc_t^y}{d\mu_t} = \frac{\lambda \mu_t}{|HB|} + (c_{t+1}^o - \omega^o) \frac{U_2''}{|HB|} \\ \frac{dc_{t+1}^o}{d\mu_t} = \frac{-\lambda}{|HB|} + (c_{t+1}^o - \omega^o) \frac{\mu_t U_1''}{|HB|} \end{cases} \tag{13}$$

Note that $\frac{\lambda \mu_t}{|HB|}$ et $\frac{-\lambda}{|HB|}$ are the slopes of the hicksian demand curves of c_t^y and c_{t+1}^o.

The expressions $(c_{t+1}^o - \omega^o) \frac{U_2''}{|HB|}$ and $(c_{t+1}^o - \omega^o) \frac{\mu_t U_1''}{|HB|}$ describe the income effects of a relative price variation.

3.2 First restriction: the elasticity price of the hicksian demands are constant

We note $E_{c_t^y|U=\bar{U}}$ as the elasticity price of the hicksian demand of a young agent, which is expressed as:

$$E_{c_t^y|U=\bar{U}} = \frac{\lambda \mu_t}{|HB|} \frac{\mu_t}{c_t^y} \tag{14}$$

using the expressions of $|HB|$ and λ we obtain:

$$E_{c_t^y|U=\bar{U}} = \frac{U_1' \mu_t^2}{-\mu_t^2 U_1'' - U_2''} \frac{1}{c_t^y} \tag{15}$$

or (see appendix 2 for the proof)

$$E_{c_t^y|U=\bar{U}} = \frac{\left(\frac{(U_2')^2}{U_2''}\right)}{\left(\frac{(U_1')^2}{U_1''} + \frac{(U_2')^2}{U_2''}\right)} \frac{-U_1'}{c_t^y U_1''} \tag{16}$$

This equation is the product of two expressions: the marginal propensity of the older to consume (see Sato [8] pp. 105 for the proof) and the inverse of the relative risk aversion coefficient of the younger. We assume first that the first term is constant. If so, then the elasticity (16) is constant if the relative risk aversion coefficient is constant. This leads to two expressions for the utility functions of a young agent:

$$\begin{cases} U_1\left(c_t^y\right) = A_1 \frac{1}{B_1+1} \left(c_t^y\right)^{B_1+1} & A_1 > 0, \quad B_1 \in \Re^- \setminus \{-1\} \\ \text{and} \\ U_1\left(c_t^y\right) = X_1 \ln\left(\Delta_1 c_t^y\right) & X_1 > 0, \quad \Delta_1 > 0 \end{cases} \tag{17}$$

Similarly, the same elasticity for an old agent is given by

$$E_{c_{t+1}^o|U=\bar{U}} = \frac{-\lambda}{|HB|} \frac{\mu_t}{c_{t+1}^o} \tag{18}$$

and using the expressions of $|HB|$ and λ we obtain:

$$E_{c_{t+1}^o|U=\bar{U}} = \frac{-\left(\frac{U_2'}{\mu_t}\right)}{-\mu_t^2 U_1'' - U_2''} \frac{\mu_t}{c_{t+1}^o} \tag{19}$$

or (see appendix 3 for the proof):

$$E_{c_{t+1}^o|U=\bar{U}} = \frac{\left(\frac{(U_1')^2}{U_1''}\right)}{\left(\frac{(U_1')^2}{U_1''} + \frac{(U_2')^2}{U_2''}\right)} \frac{U_2'}{c_{t+1}^o U_2''} \tag{20}$$

Once again, this equation is the product of two expressions: the marginal propensity of the younger to consume (see Sato [8] pp. 105 for the proof) and the inverse of relative risk aversion coefficient of the older. Assuming that the first expression is constant, then the elasticity (20) is constant if the relative risk aversion coefficient is constant. We obtain the following expressions for the utility functions of the old agent:

$$\begin{cases} U_2\left(c_{t+1}^v\right) = A_2 \frac{1}{B_2+1} \left(c_{t+1}^v\right)^{B_2+1} & A_2 > 0, \quad B_2 \in \Re^- \setminus \{-1\} \\ \text{and} \\ U_2\left(c_{t+1}^v\right) = X_2 \ln\left(\Delta_2 c_{t+1}^v\right) & X_2 > 0, \quad \Delta_2 > 0 \end{cases} \tag{21}$$

3.3   Second restriction: the Engel curves are linear

There are two steps to obtain utility functions expressions that satisfy the hypothesis of linear Engel curves.

The first one consists of obtaining a condition on the derivatives of the utility function. It can be shown (see appendix 4 for the proof) that linear Engel curves imply that the marginal propensity to consume is constant. Which leads to the following relation:

$$U_i' U_i''' - k \left(U_i'' \right)^2 = 0 \tag{22}$$

In the second step we solve this third-order differential equation and find three expressions for the utility functions. Let us define $x_1 = c_t^y$ and $x_2 = c_{t+1}^o$.

Then:

$$
\begin{cases}
k = 1 \text{ and } U_i = \frac{\alpha_i}{\beta_i} \exp \left(\frac{1}{\alpha_i} x_i + \chi_i \right) & i = 1, 2 \\
\text{where } \alpha_i < 0, \quad \beta_i < 0, \quad \chi_i \in \Re \\
k = 2 \text{ and } U_i = \frac{-\epsilon_i}{\phi_i^2} \ln \left[\gamma_i \left(\frac{\phi_i^2}{\epsilon_i} x_i + \varphi_i \right) \right] & i = 1, 2 \\
\text{where } \epsilon_i < 0, \quad \phi_i > 0, \quad (\gamma_i, \varphi_i) \in \Re^2 \\
k \in \Re \backslash \{0, 1, 2\} \text{ and } U_i = \frac{1}{\iota_i \kappa_i} \frac{1}{2-k} [\kappa_i (1 - k) (x_i - \eta_i)]^{\frac{2-k}{1-k}} & i = 1, 2 \\
\text{where } \iota_i > 0, \quad \kappa_i < 0, \quad (1 - k)(x_i - \eta_i) < 0, \quad \eta_i \in \Re
\end{cases}
\tag{23}
$$

4   The Dynamics of the Aggregate Consumption

4.1   Case 1: The dynamics when the elasticity price of the Hicksian demands are constant

As is shown in appendix 2 and 3, it is possible to derive the dynamics of aggregate consumption by considering the following combination of utility functions for the first and second periods:

$$
\begin{cases}
U_1 \left(c_t^y \right) = \Gamma_1 \ln \left(\Delta_1 c_t^y \right) & \text{where} & X_1 > 0, \quad \Delta_1 > 0 \\
U_2 \left(c_{t+1}^o \right) = \Gamma_2 \ln \left(\Delta_2 c_{t+1}^o \right) & \text{where} & X_2 > 0, \quad \Delta_2 > 0
\end{cases}
\tag{24}
$$

This selection is made according to the hypothesis that the marginal propensities of the younger and the older to consume are constant.

The consumption dynamics are studied by considering both the optimality conditions of the consumer maximisation program (7) and equation (1)

After a few transformations this leads to a first-order difference equation (see appendix 5 for the proof):

$$c_t^y = f(c_t^y) = \frac{\omega^y + (\omega^y + \omega^o)\left[\frac{\Gamma_1}{\Gamma_2}\frac{c_t^y - \omega^v}{c_t^y}\right]}{1 + \left[\frac{\Gamma_1}{\Gamma_2}\frac{c_t^y - \omega^v}{c_t^y}\right]} \qquad \text{where} \qquad \Gamma_1, \Gamma_2 > 0 \quad (25)$$

Proposition 5 *When the dynamic of the economy is governed by equation (25), the timepath of the consumption is described by a monotonic convergence towards two stationary equilibria (proof: see appendix 5).*

For a graphical illustration, the reader is referred to figures 1 ($\omega^y = \omega^o = 0.5$, $\frac{\Gamma_1}{\Gamma_2} = 0.5$, $c_0^y = 0.9$) and 2 ($\omega^y = \omega^o = 0.5$, $\frac{\Gamma_1}{\Gamma_2} = 10$, $c_0^y = 0.51$). The curves represent the timepath of the consumption. The line in each figure represent the set of equilibria corresponding to the stationnary state. In figure 1 the dynamics of consumption converge to the no trade equilibrium $(c_t)^* = 0.5$. In figure 2 the convergence is towards the second equilibrium $(c_t)^{**} = 0.909$

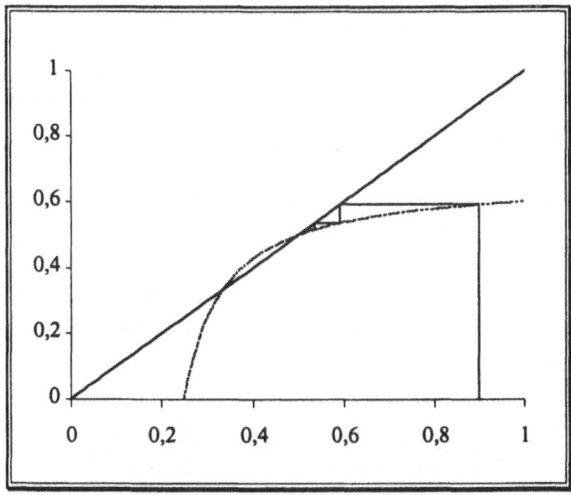

Figure 1 : No trade equilibrium

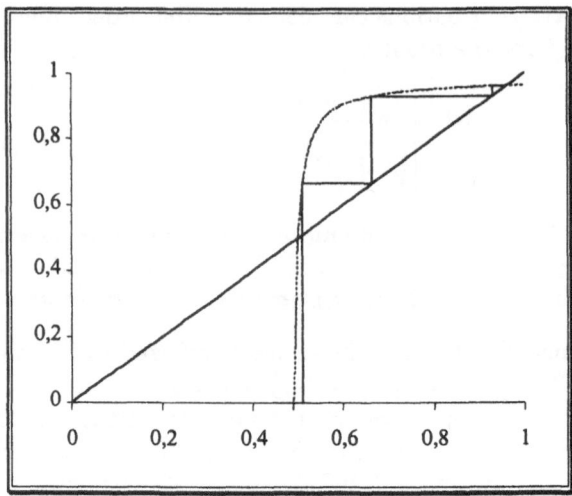

Figure 2 : Second equilibrium

4.2   CASE 2: THE DYNAMICS OF LINEAR ENGEL CURVES

Proposition 6 *Under the assumption of linearity of the Engel curves, only one specification of the utility functions is kept. Let us define $x_1 = c_t^y$ and $x_2 = c_{t+1}^o$. We have (for a proof see appendix 4):*

$$U_i(x_i) = \frac{1}{\tau_i \kappa_i} \frac{1}{2-k} \left[\kappa_i (1-k)(x_i - \eta_i) \right]^{\frac{2-k}{1-k}} + \varphi_i \qquad (26)$$
$$\text{where} \quad \kappa_i < 0, \quad \tau_i > 0, \quad \eta_\iota \in \Re \text{ and } k \in \Re \backslash \{0, 1, 2\}$$

Proposition 7 *Given the utility function represented by equation (26), the consumption dynamics is given by a first-order difference equation (for a proof see appendix 6):*

$$c_{t+1}^y = \omega^y - \left(\frac{\iota_2}{\iota_1} \right)^{\frac{1-k}{2-k}} (\omega^y - c_t^y)^{\frac{1-k}{2-k}} (c_t^y)^{\frac{1}{2-k}} \qquad (27)$$

This dynamic equation is a kind of logistic equation. Thus the timepath of the consumption will vary from an oscillatory convergence to a stationnary equilibrium to periodic cycles and chaos, given a set of values for the parameters. The figures which illustrate the consumption's timepath for different values of the parameters are represented next page. We have chosen in particular: $\omega^y = 1$, $k = 0.9$ and $\frac{\iota_2}{\iota_1}$ varying from 1.6 (convergence to stationnary equilibrium) to 1.99 (chaotic timepath).

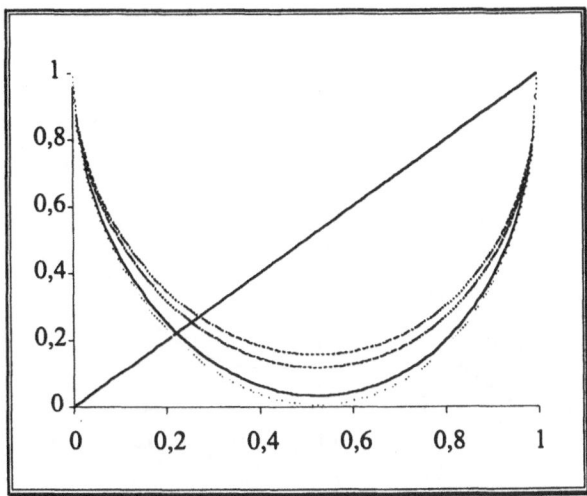

Figure 3 : Phase diagram

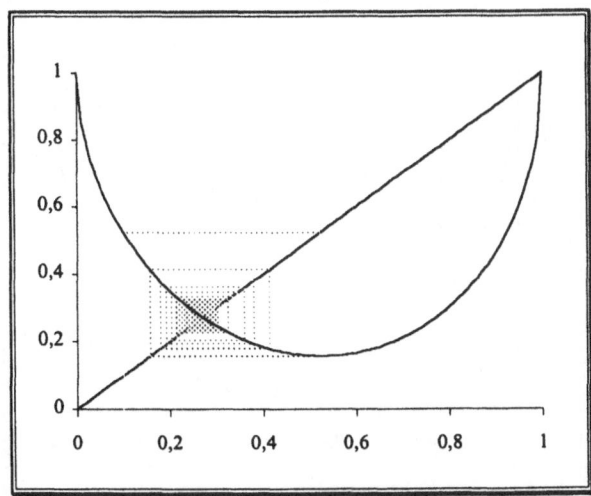

Figure 4 : Stationnary equilibrium

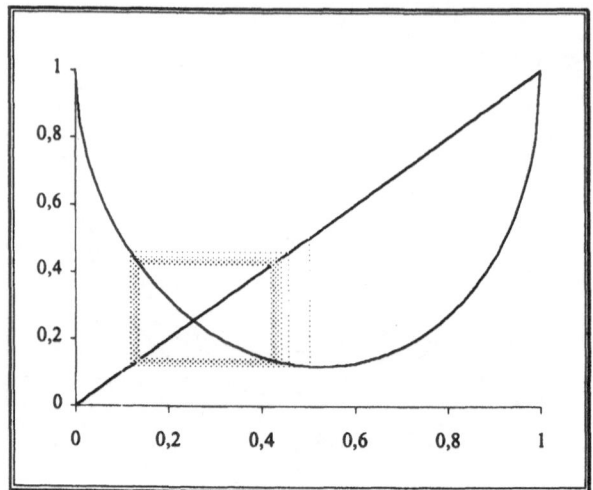

Figure 5 : Two period cycle

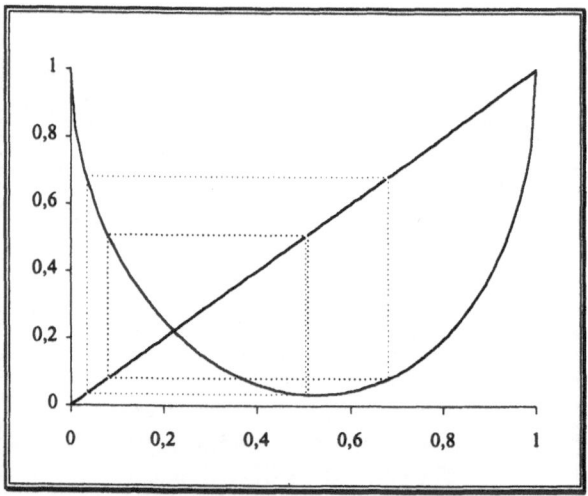

Figure 6 : Four period cycle

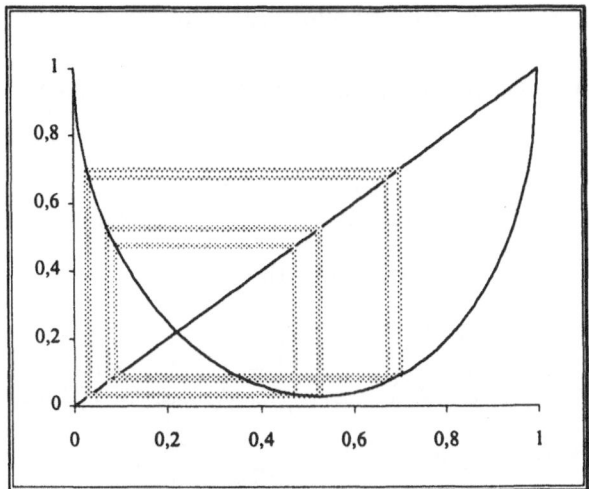

Figure 7 : Eight period cycle

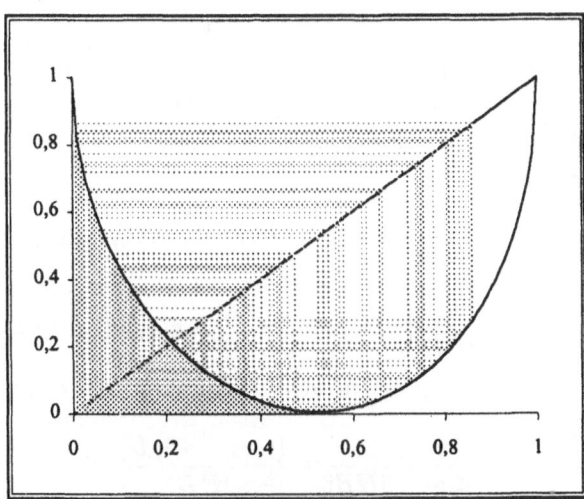

Figure 8 : Chaos

Appendix 1 :Derivation of Slutsky equations

From equation (9) pp. 91) we use a second-order differentiation:

$$\begin{cases} U_1'' dc_t^y - d\lambda = 0 \\ U_2'' dc_{t+1}^o - \mu_t d\lambda = \lambda d\mu_t \\ -dc_t^y - \mu_t dc_{t+1}^o = -d\omega^y + c_{t+1}^o d\mu_t \end{cases} \tag{28}$$

Using a matrix representation we write:

$$\begin{pmatrix} U_1'' & 0 & -1 \\ 0 & U_2'' & -\mu_t \\ -1 & -\mu_t & 0 \end{pmatrix} \begin{pmatrix} dc_t^y \\ dc_{t+1}^o \\ d\lambda \end{pmatrix} = \begin{pmatrix} 0 \\ \lambda d\mu_t \\ -d\omega^y + c_{t+1}^o d\mu_t \end{pmatrix} \tag{29}$$

The Cramer rule implies:

$$dc_t^y = \frac{1}{|HB|} \begin{vmatrix} 0 & 0 & -1 \\ \lambda d\mu_t & U_2'' & -\mu_t \\ -d\omega^y + c_{t+1}^o d\mu_t & -\mu_t & 0 \end{vmatrix} \tag{30}$$

$$dc_{t+1}^o = \frac{1}{|HB|} \begin{vmatrix} U_1'' & 0 & -1 \\ 0 & \lambda d\mu_t & -\mu_t \\ -1 & -d\omega^y + c_{t+1}^o d\mu_t & 0 \end{vmatrix} \tag{31}$$

where $|HB|$ is the determinant of the bordered hessian matrix. In a more extensive form we have:

$$dc_t^y = \frac{1}{|HB|} \left[\mu_t \lambda d\mu_t - U_2'' d\omega^y + U_2'' c_{t+1}^o d\mu_t \right] \tag{32}$$

$$dc_{t+1}^o = \frac{1}{|HB|} \left[-U_1'' \mu_t d\omega^y + U_1'' \mu_t c_{t+1}^o d\mu_t - \lambda d\mu_t \right] \tag{33}$$

We now divide each expression by $d\mu_t$ assuming that ω^y is fixed:

$$\frac{dc_t^y}{d\mu_t} = \frac{\mu_t \lambda}{|HB|} + \frac{\left(c_{t+1}^o - \omega^o \right) U_2''}{|HB|} \tag{34}$$

$$\frac{dc_{t+1}^o}{d\mu_t} = \frac{\lambda}{|HB|} + \frac{\left(c_{t+1}^o - \omega^o \right) \mu_t U_2''}{|HB|} \tag{35}$$

The first term on the right hand side is the substitution effect of a relative price variation on the quantity of goods consumed by the young agent and the old agent.

Appendix 2 : Derivation of utility functions for the first period when the price elasticity of the hicksian demand is constant

Let us define $E_{c_t^y|U=\bar{U}}$ as the price elasticity of the hicksian demand of a young agent:

$$E_{c_t^y|U=\bar{U}} = \frac{dc_t^y}{d\mu_t}\bigg|_{U=\bar{U}}\frac{\mu_t}{c_t^y} \tag{36}$$

Then,

$$E_{c_t^y|U=\bar{U}} = \frac{\lambda\mu_t}{|HB|}\frac{\mu_t}{c_t^y} \tag{37}$$

Substituting $|HB|$ and λ by their expression (note that from the first-order condition of the maximization program $\lambda = U_1' = \frac{U_2'}{\mu_t}$), we have:

$$E_{c_t^y|U=\bar{U}} = \frac{U_1'\mu_t^2}{-\mu_t^2U_1'' - U_2''}\frac{1}{c_t^y} \tag{38}$$

As $\mu_t = \frac{U_2'}{U_1'}$, we obtain:

$$E_{c_t^y|U=\bar{U}} = \frac{-U_1'\left(\frac{U_2'}{U_1'}\right)^2}{\left(\frac{U_2'}{U_1'}\right)^2 U_1'' + U_2''}\frac{1}{c_t^y} \tag{39}$$

Dividing both the denominator and the numerator by $\left(U_1'\right)^2$ the equation above becomes:

$$E_{c_t^y|U=\bar{U}} = \frac{\left(U_2'\right)^2}{\left(U_2'\right)^2 U_1'' + \left(U_1'\right)^2 U_2''}\frac{-U_1'}{c_t^y} \tag{40}$$

We factor the denominator with $U_1''U_2''$

$$E_{c_t^y|U=\bar{U}} = \frac{\left(U_2'\right)^2}{U_1''U_2''\left[\frac{(U_1')^2}{U_1''} + \frac{(U_2')^2}{U_2''}\right]}\frac{-U_1'}{c_t^y} \tag{41}$$

or,

$$E_{c_t^y|U=\bar{U}} = \frac{\frac{(U_2')^2}{U_2''}}{\left(\frac{(U_1')^2}{U_1''} + \frac{(U_2')^2}{U_2''}\right)}\frac{-U_1'}{c_t^y U_1''} \tag{42}$$

This last expression is the product of two expressions: the marginal propension of the older to consume (see Sato [8] pp. 105 for the proof) and the inverse of the young agent's relative risk aversion coefficient.

Conjecture 8 *The marginal propension to consume of the old is constant.*

Under this assumption, the elasticity (16) is constant if the relative risk aversion coefficient is constant. This leads to the following proposition.

Proposition 9 *If the marginal propensy to consume of a young is constant, then the preferences are described by the following expressions:*

$$\begin{cases} U_1\left(c_t^y\right) = A_1 \frac{1}{B_1+1} \left(c_t^y\right)^{B_1+1} & A_1 > 0, \quad B_1 \in \Re^- \backslash \{-1\} \\ \text{and} \\ U_1\left(c_t^y\right) = X_1 \ln\left(\Delta_1 c_t^y\right) & X_1 > 0, \quad \Delta_1 > 0 \end{cases} \tag{43}$$

Proof

Assume that the relative risk aversion coefficient is constant. Thus, we write:

$$\frac{U_1'}{c_t^y U_1''} = k_1 \tag{44}$$

where k_1 is a real number. From this equality, we deduce a second-order differential equation:

$$k_1 c_t^y U_1'' - U_1' = 0 \tag{45}$$

To solve this equation, we define three cases.

- $k_1 = 1$

  this implies,

$$\frac{U_1''}{U_1'} = \frac{1}{c_t^y} \tag{46}$$

If we integrate this equation, we obtain:

$$\ln\left|\beta_1 U_1'\right| = \ln\left|\alpha_1 c_t^y\right| + \chi_1 \tag{47}$$

Conjecture 10 *The constants in the equation above are defined such that: $\alpha_1, \beta_1 > 0$ and $\chi_1 = 0$*

Taking the exponential of both side yields:

$$\beta_1 U_1' = \alpha_1 c_t^y + \chi_1 \tag{48}$$

Integrating once more, we obtain:

$$U_1\left(c_t^y\right) = \frac{\alpha_1}{2\beta_1}\left(c_t^y\right)^2 + \kappa_1 \tag{49}$$

Unfortunately, this function is not strictly concave:

$$\begin{cases} U_1' = \frac{\alpha_1}{\beta_1} c_t^y > 0 & \text{since} \quad \alpha_1, \beta_1 > 0 \\ \text{and} \\ U_1'' = \frac{\alpha_1}{\beta_1} > 0 & \text{since} \quad \alpha_1, \beta_1 > 0 \end{cases} \tag{50}$$

We thus eliminate this specification from our study.

- $k_1 = -1$

  this implies,

$$\frac{U_1''}{U_1'} = \frac{-1}{c_t^y} \tag{51}$$

Integrating this equation leads to:

$$\ln \left| \beta_1 U_1' \right| = -\ln \left| \alpha_1 c_t^y \right| + \chi_1 \tag{52}$$

Conjecture 11 *The constants are assumed to satisfy the following conditions:* $\alpha_1, \beta_1 > 0$ *and* $\chi_1 = 0$

Taking the exponential of both sides yields,

$$\beta_1 U_1' = \frac{1}{\alpha_1 c_t^y} \tag{53}$$

We integrate once again:

$$U_1 = \frac{1}{\alpha_1 \beta_1} \ln \left| \varpi_1 c_t^y \right| + \epsilon_1 \tag{54}$$

Conjecture 12 *The constants are assumed to satisfy the following conditions:* $\varpi_1 > 0$ *and* $\epsilon_1 = 0$.

We thus obtain a second specification of the utility function for a young agent:

$$U_1 = \frac{1}{\alpha_1 \beta_1} \ln \left| \varpi_1 c_t^y \right| \tag{55}$$

This function is monotone and strictly concave:

$$\begin{cases} U_1' = \frac{1}{\alpha_1 \beta_1} \frac{1}{c_t^y} > 0 & \text{since} & \alpha_1, \beta_1 > 0 \\ \text{and} & & \\ U_1'' = \frac{-1}{\alpha_1 \beta_1} \frac{1}{(c_t^y)^2} < 0 & \text{since} & \alpha_1, \beta_1 > 0 \end{cases} \tag{56}$$

- $k_1 \neq 1$ and $k_1 \neq -1$

  This implies

$$k_1 \frac{U_1''}{U_1'} = \frac{1}{c_t^y} \tag{57}$$

We integrate and obtain:

$$k_1 \ln \left| \gamma_1 U_1' \right| = \ln \left| \delta_1 c_t^y \right| + \chi_1 \tag{58}$$

Conjecture 13 *The constants are assumed to satisfy the following conditions:* $(\delta_1, \gamma_1) > 0$ *and* $\chi_1 = 0$

Taking the exponential of both sides yields:

$$\gamma_1 U_1' = \delta_1 c_t^y \tag{59}$$

or,

$$U_1' = \frac{(\delta_1)^{1/k_1}}{\gamma_1} (c_t^y)^{1/k_1} \tag{60}$$

We integrate once more and obtain a third specification of the utility function of the young agent:

$$U_1 = \frac{(\delta_1)^{1/k_1}}{\gamma_1} \frac{1}{1/k_1 + 1} (c_t^y)^{1/k_1 + 1} \tag{61}$$

This function is also monotone and strictly concave:

$$\begin{cases} U_1' = \frac{(\delta_1)^{1/k_1}}{\gamma_1} (c_t^y)^{1/k_1} > 0 & \text{since} \quad \delta_1, \gamma_1 > 0 \\ \text{and} \\ U_1'' = \frac{(\delta_1)^{1/k_1}}{\gamma_1} \frac{1}{k_1} (c_t^y)^{1/k_1 - 1} < 0 & \text{if} \quad k_1 < 0 \end{cases} \tag{62}$$

Appendix 3 : Derivation of utility functions for the second period when the price elasticity of the hicksian demand is constant

Let us define $E_{c^o_{t+1}|U=\bar{U}}$ as the price elasticity of the hicksian demand of an old agent:

$$E_{c^o_{t+1}|U=\bar{U}} = \frac{dc^o_{t+1}}{d\mu_t |U=\bar{U}} \frac{\mu_t}{c^o_{t+1}} \tag{63}$$

Then,

$$E_{c^o_{t+1}|U=\bar{U}} = \frac{-\lambda}{|HB|} \frac{\mu_t}{c^o_{t+1}} \tag{64}$$

Substituting $|HB|$ and λ by their expression (note that from the first-order condition of the maximisation program $\lambda = U'_1 = \frac{U'_2}{\mu_t}$), we have:

$$E_{c^o_{t+1}|U=\bar{U}} = \frac{U'_1 \mu_t}{\mu_t^2 U''_1 + U''_2} \frac{1}{c^o_{t+1}} \tag{65}$$

As $\mu_t = \frac{U'_2}{U'_1}$, we obtain:

$$E_{c^o_{t+1}|U=\bar{U}} = \frac{U'_2}{\left(\frac{U'_2}{U'_1}\right)^2 U''_1 + U''_2} \frac{1}{c^o_{t+1}} \tag{66}$$

We factor the denominator with $U''_1 U'''_2$:

$$E_{c^o_{t+1}|U=\bar{U}} = \frac{\left(U'_1\right)^2}{U''_1 U''_2 \left[\frac{(U'_1)^2}{U''_1} + \frac{(U'_2)^2}{U''_2}\right]} \frac{U'_2}{c^o_{t+1}} \tag{67}$$

or,

$$E_{c^o_{t+1}|U=\bar{U}} = \frac{\frac{\left(U'_1\right)^2}{U''_1}}{\left(\frac{(U'_1)^2}{U''_1} + \frac{(U'_2)^2}{U''_2}\right)} \frac{U'_2}{c^o_{t+1} U''_2} \tag{68}$$

This last expression is the product of two expressions: the marginal propensity of the younger to consume (see Sato [8] pp. 105 for the proof) and the inverse of the old agent's relative risk aversion coefficient.

Conjecture 14 *The marginal propensity of the young agent to consume is constant.*

With this assumption, the elasticity (16) is constant if the relative risk aversion coefficient is constant. This leads to the following proposition:

Proposition 15 *If the marginal propensity of an old agent to consume is constant, then the preferences are described by the following expressions:*

$$\begin{cases} U_2\left(c_{t+1}^o\right) = A_2 \frac{1}{B_2+1}\left(c_{t+1}^o\right)^{B_2+1} & A_2 > 0, \quad B_2 \in \Re^- \backslash \{-1\} \\ and \\ U_2\left(c_{t+1}^o\right) = X_2 \ln\left(\Delta_2 c_{t+1}^o\right) & X_2 > 0, \quad \Delta_2 > 0 \end{cases} \tag{69}$$

Proof : see appendix 2.

Appendix 4 : Derivation of utility functions when the Engel curves are linear.

Conjecture 16 *The Engel Curves are linear.*

Proposition 17 *Under this assumption, we derive three specifications of the utility functions. Let us define $x_1 = c_t^y$ and $x_2 = c_{t+1}^o$. We have:*

$$
\begin{cases}
U_i(x_i) = \frac{\alpha_i}{\beta_i} \exp\left(\frac{1}{\alpha_i} x_i + \chi_i\right) + \delta_i \\
\text{where} \quad \alpha_i, \beta_i < 0 \text{ et } \chi_i \in \Re \\
U_i(x_i) = \frac{-\epsilon_i}{(\phi_i)^2} \ln\left[\gamma_i \left(\frac{(\phi_i)^2}{\epsilon_i}\right) x_i + \varphi_i\right] \\
\text{where} \quad \epsilon_i < 0, \quad \phi_i > 0 \text{ and } (\gamma_i, \varphi_i) \in \Re^2 \\
U_i(x_i) = \frac{1}{\tau_i \kappa_i} \frac{1}{2-k} \left[\kappa_i (1-k)(x_i - \eta_i)\right]^{\frac{2-k}{1-k}} + \varphi_i \\
\text{where} \quad \kappa_i < 0, \quad \tau_i > 0, \quad \eta_i \in \Re \text{ et } k \in \Re \setminus \{0, 1, 2\}
\end{cases}
\tag{70}
$$

Proof

Two steps are needed to prove this proposition.

1. Linearity condition of the Engel Curves.

   The linearity of the Engel Curves implies that the marginal propensities of the young and the old agents' to consume are constants.

   Let us define pmc_i as the marginal propensity of the young agent ($i = 1$) and the old agent ($i = 2$) to consume. The linearity condition of the Engel curves implies:

   $$
   \frac{\partial pmc_i}{\partial(\omega^y + \mu_t \omega^o)} = 0
   \tag{71}
   $$

   This condition is true iff $\frac{U_i' U_i'''}{(U_i'')^2}$ is constant whith any values of c_t^y and c_{t+1}^o (see Sato [8], pp. 117 for the proof). Thus:

   $$
   \frac{U_i' U_i'''}{(U_i'')^2} = k, \quad \text{for } i = 1, 2, \text{ where } k \in \Re \setminus \{0\}
   \tag{72}
   $$

   Note that the expression above leads to a third-order differential equation:

   $$
   U_i' U_i''' - k \left(U_i''\right)^2 = 0
   \tag{73}
   $$

2. Derivation of the utility functions.

   Solving this equation yields three specifications depending upon the value of the parameter k.

(a) $k = 1$

This implies,

$$U_i' U_i''' = \left(U_i''\right)^2 \tag{74}$$

or,

$$\frac{\left(U_i''\right)^2 - U_i' U_i'''}{\left(U_i''\right)^2} = 0 \tag{75}$$

Integrating these equation gives:

$$\frac{U_i'}{U_i''} = \alpha_i \quad \text{or} \quad \frac{U_i''}{U_i'} = \frac{1}{\alpha_i} \tag{76}$$

where α_i is a real number. Integrating once again, we obtain:

$$\ln \left| \beta_i U_i' \right| = \frac{1}{\alpha_i} x_i + \chi_i \tag{77}$$

Conjecture 18 *We assume that: $\beta_i > 0$*

Under this assumption, we take the exponential of both sides of the equality and obtain:

$$U_i'(x_i) = \frac{1}{\beta_i} \exp \left(\frac{1}{\alpha_i} x_i + \chi_i \right) \tag{78}$$

A new integration gives a first specification of the utility function:

$$U_i(x_i) = \frac{\alpha_i}{\beta_i} \exp \left(\frac{1}{\alpha_i} x_i + \chi_i \right) + \delta_i \tag{79}$$

where δ_i is a constant which we assume to be equal to zero.
This function is monotone and strictly concave:

$$\begin{cases} U_i'(x_i) = \frac{1}{\beta_i} \exp \left(\frac{1}{\alpha_i} x_i + \chi_i \right) > 0 \quad \text{since} \quad \beta_i > 0 \\ \text{and} \\ U_i''(x_i) = \frac{1}{\beta_i} \frac{1}{\alpha_i} \exp \left(\frac{1}{\alpha_i} x_i + \chi_i \right) < 0 \quad \text{if} \quad \alpha_i < 0 \end{cases} \tag{80}$$

(b) $k = 2$

This implies,

$$U_i' U_i''' = 2 \left(U_i''\right)^2 \tag{81}$$

or,

$$\frac{U_i'''}{U_i''} = \frac{2U_i''}{U_i'} \tag{82}$$

We integrate both sides and obtain:

$$\ln \left| \epsilon_i U_i'' \right| = 2 \ln \left| \phi_i U_i' \right| + \zeta_i \tag{83}$$

Conjecture 19 *We assume that:* $\epsilon_i < 0, \quad \phi_i > 0, \quad \zeta_i = 0$

One thus has:

$$\ln\left(\epsilon_i U_i''\right) = 2\ln\left(\phi_i U_i'\right) \tag{84}$$

Taking the exponential of both sides gives:

$$U_i'' = \frac{1}{\epsilon_i}\left(\phi_i U_i'\right)^2 \tag{85}$$

or,

$$\left(U_i'\right)^{-2} U_i'' = \frac{(\phi_i)^2}{\epsilon_i} \tag{86}$$

Integrating, we obtain:

$$\left(U_i'\right)^{-1} = \frac{(\phi_i)^2}{\epsilon_i} x_i + \varphi_i \tag{87}$$

or,

$$U_i' = \frac{-1}{\frac{(\phi_i)^2}{\epsilon_i} x_i + \varphi_i} \tag{88}$$

Integrating once again, we obtain a second specification of the utility function:

$$U_i\left(x_i\right) = \frac{-\epsilon_i}{(\phi_i)^2} \ln\left[\gamma_i\left(\frac{(\phi_i)^2}{\epsilon_i} x_i + \varphi_i\right)\right] + \xi_i \tag{89}$$

where we suppose that $\xi_i = 0$ and $\gamma_i\left(\frac{(\phi_i)^2}{\epsilon_i} x_i + \varphi_i\right) > 0$. This expression, which is a Stone-Geary utility function, is monotone and strictly concave:

$$\begin{cases} U_i'\left(x_i\right) = \frac{-1}{\frac{(\phi_i)^2}{\epsilon_i} x_i + \varphi_i} > 0 \quad \text{if} \quad \frac{(\phi_i)^2}{\epsilon_i} x_i + \varphi_i < 0 \\ \text{and} \\ U_i''\left(x_i\right) = \frac{\frac{(\phi_i)^2}{\epsilon_i}}{\left(\frac{(\phi_i)^2}{\epsilon_i} x_i + \varphi_i\right)^2} < 0 \quad \text{since} \quad \epsilon_i < 0 \end{cases} \tag{90}$$

(c) $k \neq 1$ and $k \neq 2$
This implies,

$$\frac{U_i' U_i'''}{\left(U_i''\right)^2} = k \tag{91}$$

or,

$$-U_i' U_i''' = -(k + 1 - 1)\left(U_i''\right)^2 \tag{92}$$

This yields,

$$\left(U_i''\right)^2 - U_i' U_i''' = -(k-1)\left(U_i''\right)^2 \tag{93}$$

which can be written as,

$$\frac{\left(U_i''\right)^2 - U_i' U_i'''}{\left(U_i''\right)^2} = -(k-1) \tag{94}$$

We integrate both sides and obtain:

$$\frac{U_i'}{U_i''} = (1-k)(x_i - \eta_i) \tag{95}$$

where $(1-k)(-\eta_i)$ is a constant. We can also write:

$$\frac{U_i''}{U_i'} = \frac{1}{(1-k)(x_i - \eta_i)} \tag{96}$$

Integrating this expression leads to:

$$\ln\left|\iota_i U_i'\right| = \frac{1}{(1-k)} \ln|\kappa_i (1-k)(x_i - \eta_i)| + \upsilon_i \tag{97}$$

where ι_i, κ_i and υ_i are real numbers.

Conjecture 20 *We assume :* $\kappa_i (1-k)(x_i - \eta_i) > 0$, $\iota_i > 0$, $\upsilon_i = 0$

and thus write:

$$\ln\left(\epsilon_i U_i''\right) = 2\ln\left(\phi_i U_i'\right) \tag{98}$$

Taking the exponential of both sides we obtain:

$$U_i' = \frac{1}{\iota_i}\left[\kappa_i (1-k)(x_i - \eta_i)\right]^{\frac{1}{1-k}} \tag{99}$$

Integrating once more we derive a third specification of the utility function:

$$U_i(x_i) = \frac{1}{\iota_i \kappa_i}\frac{1}{2-k}\left[\kappa_i (1-k)(x_i - \eta_i)\right]^{\frac{2-k}{1-k}} + \tau_i \tag{100}$$

where τ_i is assumed to be equal to zero.

This function is also monotone and strictly concave:

$$\begin{cases} U_i'(x_i) = \frac{1}{\iota_i}\left[\kappa_i (1-k)(x_i - \eta_i)\right]^{\frac{1}{1-k}} > 0 \\ \quad \text{if} \quad \iota_i > 0 \\ \text{and} \\ U_i''(x_i) = \frac{\kappa_i}{\iota_i}\left[\kappa_i (1-k)(x_i - \eta_i)\right]^{\frac{k}{1-k}} < 0 \\ \text{if} \quad \kappa_i < 0 \text{ and if } (1-k)(x_i - \eta_i) < 0 \end{cases} \tag{101}$$

Appendix 5 : The dynamics of the aggregate consumption when the elasticity price of the hicksian demands is constant

From appendix 2 and 3, we use the following specifications of the utility functions:

$$\begin{cases} U_1\left(c_t^y\right) = \Gamma_1 \ln\left(\Delta_1 c_t^y\right) & \text{where} \quad X_1 > 0, \quad \Delta_1 > 0 \\ \text{and} \\ U_2\left(c_{t+1}^o\right) = \Gamma_2 \ln\left(\Delta_2 c_{t+1}^o\right) & \text{where} \quad X_2 > 0, \quad \Delta_2 > 0 \end{cases} \tag{102}$$

The optimality condition can be written as follows:

$$\left(c_t^y - \omega^y\right) U_1' = \left(\omega^o - c_{t+1}^o\right) U_2' \tag{103}$$

The equilibrium on the good market implies:

$$c_{t+1}^v = \omega^o + \omega^y - c_t^y \tag{104}$$

The derivatives of the utility functions yield:

$$U_1' = \frac{\Gamma_1}{c_t^y} \quad \text{and} \quad U_2' = \frac{\Gamma_2}{c_{t+1}^o} \tag{105}$$

Combining all these expressions, we obtain:

$$\left(c_t^y - \omega^y\right) \frac{\Gamma_1}{c_t^y} = \left(c_t^y - \omega^y\right) \frac{\Gamma_2}{\omega^o + \omega^y - c_t^y} \tag{106}$$

Proposition 21 *The dynamics of aggregate consumption of the young is described by a first-order difference equation.*

Proof
From the equation (106), we reach:

$$c_t^y = f\left(c_t^y\right) = \frac{\omega^y + \left(\omega^y + \omega^o\right)\left[\frac{\Gamma_1}{\Gamma_2}\frac{c_t^y - \omega^v}{c_t^y}\right]}{1 + \left[\frac{\Gamma_1}{\Gamma_2}\frac{c_t^y - \omega^v}{c_t^y}\right]} \tag{107}$$

Conjecture 22 *We assume that :* $\omega^y + \omega^o = 1$ *and* $\left(\omega^y, \omega^o\right) \neq 1$.

Thus, the difference equation is rewritten:

$$c_t^y = f\left(c_t^y\right) = \frac{\omega^y + \left[\frac{\Gamma_1}{\Gamma_2}\frac{c_t^y - \omega^v}{c_t^y}\right]}{1 + \left[\frac{\Gamma_1}{\Gamma_2}\frac{c_t^y - \omega^v}{c_t^y}\right]} \tag{108}$$

The first-order derivative of $f\left(c_t^y\right)$ provides some information on the timepath of the consumption (monotonic convergence, oscillatory evolutions, chaotic dynamics).

1. The dynamic properties of $f(c_t^y)$

Let us define f' as the first-order derivative of f:

$$f'(c_t^y) = \frac{\frac{\Gamma_1}{\Gamma_2}\frac{\omega^y-(\omega^y)^2}{(c_t^y)^2}}{\left[1+\left[\frac{\Gamma_1}{\Gamma_2}\frac{c_t^y-\omega^y}{c_t^y}\right]\right]^2} > 0 \quad \text{since} \quad \omega^y > (\omega^y)^2 \qquad (109)$$

This derivative is stricly positive thus f is an increasing function of c_t^y .

2. The domain of definition of $f(c_t^y)$

Since we assume that $\omega^y + \omega^o \leq 1$, $c_t^y \leq 1$. Further, $f(c_t^y)$ exists for values of c_t^y in the interval $\frac{\Gamma_1}{2\Gamma_1+\Gamma_2} \leq c_t^y \leq 1$.

Indeed:

$$\begin{cases} \frac{\Gamma_1\omega^y}{\Gamma_1+\Gamma_2} \leq c_t^y \leq \frac{\Gamma_1}{2\Gamma_1+\Gamma_2} \Rightarrow f(c_t^y) < 0 \\ \text{and} \\ c_t^y \leq \frac{\Gamma_1\omega^y}{\Gamma_1+\Gamma_2} \Rightarrow f(c_t^y) > 1 \end{cases} \qquad (110)$$

Thus,

$$f : \left[\frac{\Gamma_1}{2\Gamma_1+\Gamma_2}, 1\right] \qquad (111)$$

3. Steady states equilibria

A steady state equilibrium must satisfy the following equation:

$$c_t^y = \frac{\omega^y + \left[\frac{\Gamma_1}{\Gamma_2}\frac{c_t^y-\omega^y}{c_t^y}\right]}{1+\left[\frac{\Gamma_1}{\Gamma_2}\frac{c_t^y-\omega^y}{c_t^y}\right]} \qquad (112)$$

Since $c_t^y \neq 0$ we can multiply both the numerator and denominator by c_t^y:

$$c_t^y = \frac{c_t^y\omega^y + \frac{\Gamma_1}{\Gamma_2}(c_t^y-\omega^y)}{c_t^y + \frac{\Gamma_1}{\Gamma_2}(c_t^y-\omega^y)} \qquad (113)$$

We obtain,

$$c_t^y\left[c_t^y + \frac{\Gamma_1}{\Gamma_2}(c_t^y-\omega^y)\right] = c_t^y\omega^y + \frac{\Gamma_1}{\Gamma_2}(c_t^y-\omega^y) \qquad (114)$$

which leads to a second-order polynomial:

$$\left(1+\frac{\Gamma_1}{\Gamma_2}\right)(c_t^y)^2 - \left[\frac{\Gamma_1}{\Gamma_2}+\frac{\Gamma_1}{\Gamma_2}\omega^y+\omega^y\right]c_t^y + \frac{\Gamma_1}{\Gamma_2}\omega^y = 0 \qquad (115)$$

A first solution is $(c_t^y)^* = \omega^y$, which corresponds to the no-trade equilibrium.

A second solution is:

$$(c_t^y - \omega^y)\left[\left(1 + \frac{\Gamma_1}{\Gamma_2}\right)c_t^y - \frac{\Gamma_1}{\Gamma_2}\right] = 0 \tag{116}$$

from which we deduce:

$$(c_t^y)^* = \omega^y \text{ and } (c_t^y)^{**} = \frac{\Gamma_1}{\Gamma_1 + \Gamma_2} \tag{117}$$

4. Stability of steady states equilibria

To study the stability of the steady states equilibria, we need to evaluate the first-order derivative when $c_t^y = (c_t^y)^*$ and $c_t^y = (c_t^y)^{**}$. An equilibrium value x^* is stable if $\left|f'(x^*)\right| < 1$.

- We first check this condition for the no-trade equilibrium $(c_t^y)^* = \omega^y$

$$f'((c_t^y)^*) = \frac{\frac{\Gamma_1}{\Gamma_2}\frac{\omega^y - (\omega^y)^2}{(\omega^y)^2}}{\left[1 + \left[\frac{\Gamma_1}{\Gamma_2}\frac{\omega^y - \omega^y}{\omega^y}\right]\right]^2} = \frac{\Gamma_1}{\Gamma_2}\frac{\omega^y - (\omega^y)^2}{(\omega^y)^2} \tag{118}$$

We deduce:

$$f'((c_t^y)^*) = \frac{\Gamma_1}{\Gamma_2}\frac{1 - \omega^y}{\omega^y} \tag{119}$$

thus,

$$-1 < \frac{\Gamma_1}{\Gamma_2}\frac{1 - \omega^y}{\omega^y} < 1 \tag{120}$$

The term between the two inequalities is strictly positive. The stability condition corresponding to the no-trade situation is:

$$\frac{\Gamma_1}{\Gamma_2} < \frac{\omega^y}{1 - \omega^y} \tag{121}$$

As an illustration, we choose $\omega^y = 0.5 = (c_t^y)^*$. The inequality above becomes $\frac{\Gamma_1}{\Gamma_2} < 1$. The convergence towards the steady state equilibrium is monotonic $f'((c_t^y)^*) = 0.5 > 0$ (this is true under the condition $\frac{\Gamma_1}{\Gamma_2} = 0.5$).

- We now study the stability of the second steady state equilibrium $(c_t^y)^{**} = \frac{\Gamma_1}{\Gamma_1 + \Gamma_2}$. As f is strictly positive, we have:

$$\frac{\Gamma_1}{\Gamma_2} < \frac{\omega^y}{1 - \omega^y} \tag{122}$$

and one can easily see that the convergence is monotonic.

Appendix 6 : The dynamics of the aggregate consumption when the Engel curves are linear.

We derive three cases.

1. Case 1

   Let us define $x_1 = c_t^y$ and $x_2 = c_{t+1}^o$. The expression of the first utility function is:

   $$U_i\left(x_i\right) = \frac{\alpha_i}{\beta_i}\exp\left(\frac{1}{\alpha_i}x_i + \chi_i\right) \text{ where } i = 1, 2, \ \alpha_i, \beta_i < 0 \text{ and } \chi_i \in \Re \tag{123}$$

   We thus have:

   $$U_i'\left(x_i\right) = \frac{1}{\beta_i}\exp\left(\frac{1}{\alpha_i}x_i + \chi_i\right) \tag{124}$$

   which leads to the following expression of the consumption dynamics:

   $$\left(c_t^y - \omega^y\right)\frac{1}{\beta_1}e^{\left(\frac{1}{\alpha_1}c_t^y + \chi_1\right)} = \left(c_t^y - \omega^y\right)\frac{1}{\beta_2}e^{\left(\frac{1}{\alpha_2}\left(\omega^y + \omega^o - c_t^y\right) + \chi_2\right)} \tag{125}$$

   Unfortunately, it is impossible to find the inverse of the right hand side of this equality. Therefore, there is not a unique relation between c_t^y and c_t^y.

2. Case 2

   We consider the following utility function:

   $$U_i\left(x_i\right) = \frac{-\epsilon_i}{(\phi_i)^2}\ln\left[\gamma_i\left(\frac{(\phi_i)^2}{\epsilon_i}\right)x_i + \varphi_i\right] \tag{126}$$
   $$\text{where } \epsilon_i < 0, \quad \phi_i > 0 \text{ and } (\gamma_i, \varphi_i) \in \Re^2$$

   The first-order derivative is written as:

   $$U_i'\left(x_i\right) = \frac{-1}{\frac{(\phi_i)^2}{\epsilon_i}x_i + \varphi_i} \tag{127}$$

   Conjecture 23 *We assume that:* $\varphi_i = 0$

   Proposition 24 *The dynamics of the consumption is given by a first-order difference equation.*

   Proof

   The dynamics of the consumption is given by (let $\frac{(\phi_i)^2}{\epsilon_i} = \Gamma_i$ $i = 1, 2$):

   $$\left(c_t^y - \omega^y\right)\frac{1}{\Gamma_1 c_t^y} = \left(c_t^y - \omega^y\right)\frac{1}{\Gamma_1\left(\omega^y + \omega^o - c_t^y\right)} \tag{128}$$

so,

$$c_t^y = f(c_t^y) = \frac{\omega^y + (\omega^y + \omega^o) \left[\frac{\Gamma_1}{\Gamma_2} \frac{c_t^y - \omega^y}{c_t^y} \right]}{1 + \left[\frac{\Gamma_1}{\Gamma_2} \frac{c_t^y - \omega^y}{c_t^y} \right]} \quad \text{where} \quad \Gamma_1, \Gamma_2 > 0 \quad (129)$$

3. Case 3

We consider the following utility function:

$$\begin{cases} U_i(x_i) = \frac{1}{\tau_i \kappa_i} \frac{1}{2-k} \left[\kappa_i (1-k)(x_i - \eta_i) \right]^{\frac{2-k}{1-k}} + \varphi_i \\ \text{where} \quad \kappa_i < 0, \quad \tau_i > 0, \quad \eta_\iota \in \Re \text{ et } k \in \Re \setminus \{0, 1, 2\} \end{cases} \quad (130)$$

The first-order derivative is written as:

$$U_i' = \frac{1}{\iota_i} \left[\kappa_i (1-k)(x_i - \eta_i) \right]^{\frac{1}{1-k}} \quad (131)$$

Conjecture 25 $\omega^o = 0$

Conjecture 26 $\kappa_i = -1(1-k)$ and $\eta_i = 0$

Proposition 27 *Under these assumptions the dynamics of the consumption is given by a non-linear first-order difference equation.*

Proof

The consumption dynamics is:

$$(c_t^y - \omega^y) \frac{1}{\iota_1} \left[(k-1) c_t^y \right]^{\frac{1}{1-k}} = (c_t^y - \omega^y) \frac{1}{\iota_2} \left[(k-1)(\omega^y - c_t^y) \right]^{\frac{1}{1-k}} \quad (132)$$

from which we deduce:

$$\frac{\iota_2}{\iota_1} (c_t^y - \omega^y)(c_t^y)^{\frac{1}{1-k}} = (c_t^y - \omega^y)(\omega^y - c_t^y)^{\frac{1}{1-k}} \quad (133)$$

or,

$$\frac{\iota_2}{\iota_1} (\omega^y - c_t^y)(c_t^y)^{\frac{1}{1-k}} = (\omega^y - c_t^y)^{\frac{2-k}{1-k}} \quad (134)$$

and then,

$$c_t^y = \omega^y - \left(\frac{\iota_2}{\iota_1} \right)^{\frac{1-k}{2-k}} (\omega^y - c_t^y)^{\frac{1-k}{2-k}} (c_t^y)^{\frac{1}{2-k}} \quad (135)$$

This recurrence equation is a kind of logistic equation.

References

[1] CARROL C. [1992], "The Buffer-Stock Theory of Saving: Some Macroeconomic Evidence", *Brookings Papers on Economic Activity*, 2, 61-135.

[2] CAROLL C., SUMMERS L. [1991], *Consumption Growth Parallels Income Growth: Some New Evidence*, University of Chicago Press.

[3] DUBOIS E., BONNET X. [1996], "Peut-on comprendre la hausse imprévue du taux d'épargne des ménages depuis 1990?", *Economie et Prévision*, 121, 39-58.

[4] DYNAN K.E. [1993], "How Prudent are Consumers?, *Journal of Political Economy*, 101, 1104-1113.

[5] DREZE J., MODIGLIANI F. [1972], "Consumption Decisions under Uncertainty", *Journal of Economic Theory*, 5, 308-335.

[6] GALE D. [1973], "Pure Exchange Equilibrium of Dynamic Economic Models", *Journal of Economic Theory*, 6, 12-36.

[7] GRANDMONT J.M. [1985], "On Endogenous Competitive Business Cycles", *Econometrica*, 53, 995-1045.

[8] SATO K. [1972], "Additive Utility Functions with Double-Log Consumer Demand Functions", *Journal of Political Economy*, 102-124.

[9] SHAPIRO M.D. [1993], "The Permanent Income Hypothesis and the Interest Rate: Some Evidence from Panel Data", *Economics Letters*, 97, 305-346.

Pay-as-You-Go System under Permanent Business Cycle

Michel BOTOMAZAVA and Vincent TOUZÉ[1]

1 Introduction

The literature dealing with overlapping generations economy, so called because of its assumed demographic structure, is vast. OLG economy was introduced by Allais [1947] followed by Samuelson [1958]. The three main application fields are notably pay-as-you-go systems, endogenous business cycles and intergenerational equity theories. The purpose of this paper is to combine these three types of applications to study the pay-as-you-go system effects on the economic cycle in terms of intergenerational justice.

The first field concerns pay-as-you-go system. Two results coexist and generally deal with a steady state study: the pay-as-you-go system is or is not welfare improving. When Diamond [1965]'s OLG structure is used (Breyer and Straub [1993]), the pay-as-you-go system is not Pareto improving when the capital yield is higher than the social security yield i.e. population growth rate plus wages growth rate. However, in using Barro [1974]'s model with bequest motive, one can prove individuals offset the pay-as-you-go system effects by an offsetting bequest such that the transfers between generations are unaffected. Moreover, Zhang [1995] showed that when the population rate is endogenous, a pay-as-you-go program may stimulate growth by reducing fertility and increasing the ratio of human capital investment per child.

Second, Gale [1973] was the first to find an example of an OLG economy with two-period cycles. Benhabib and Day [1982], Grandmont [1985], Reichlin [1986] and Jullien [1988] studied the emergence of erratic or cyclic dynamics in the context of OLG models. More recently, Galor [1992] investigated OLG model dynamics in a two-sector structure. The main orientation of this research is to find and to characterize cyclic or chaotic

[1]We are grateful to Catherine TOUZÉ for helpful comments and to Antonio MELE for a pertinent correction. Of course, any remaining errors are ours.

time paths.

Third, generational equity time paths with OLG models were studied by Phelps and Riley [1978] and Rodriguez [1981]. These authors investigated the properties of "maximin" growth as meant by Rawls [1971]: so a fair society would program its taxes and resulting stocks of capital and national debt so as to maximize the lifetime utility of generations with having the lowest welfare level. In these studies, the growth is stable and convergent. A case of generational equity where the growth is cyclic has been studied by Reichlin [1986]. He searches and finds a fiscal policy rule which implies equal welfare for all generations when it is announced.

In this paper, we use these three economic concepts to study the effects of change in a pay-as-you-go system. Our pay-as-you-go system is particular. No direct pension is distributed but old age public expenditures are financed by a payroll tax. This public good cannot perfectly substitute for private good[2].

In a first part, we present Reichlin [1986]'s model when this pay-as-you-go system is introduced. We find a steady state capital general explicit form related to exogenous economic parameters.

Then, in the second part, the conditions for the existence and the stability of the cycles are given. As we want to obtain a permanent cycle, we must accept to have a parameter which varies when the payroll tax changes. This condition is necessary but we propose and then show that it could be the result of a special maximization behavior. Consequences are interesting since we can study the pay-as-you-go effects on an efficiency criterium such as the steady state level and on intergenerational justice criteria such as a welfare range gap (the egalitarian criterium) or such as the lowest welfare level (the maximin/Rawlsian criterium).

In the last part, we define an example where functional forms are known. By using computing simulations, we find there are cases where the lowest welfare level, the intergenerational inequality and the steady state level can grow up if the payroll tax rate rises. These results show two interesting conflicts: a conflict between efficiency (steady state) and equity (welfare range) criteria and a conflict between two equity criteria (welfare range and lowest welfare level).

---

[2]If the public expenditures consumption can pertfectly substitute for private good, i.e. if one additional unit of old age public expenditures gives the same additional welfare as one additional unit of private good consumption, the pay-as-you-go system is an unfunded pension system. In our numerical example, old age public good consumption can be perfectly substitute for private good consumption.

2 Pay-as-you-go system in Reichlin's model

The conditions of existence and stability for a production and overlapping generations cyclical economy were given by Reichlin [1986]. The economy consists of two types of agents: two generations of households (young and old) and a firm. In this paper the new feature is the introduction of a payroll tax rate denoted τ to finance old-age public expenditures.

2.1 THE MODEL

Each generation of households lives for two periods. The population is constant. During each period t, the younger generation offers l_t units of labor and saves its net wages and the elder consumes its accumulated saving and old-age public good. Welfare is a sum of private good consumption, old-age public good consumption and leisure utilities respectively $u(c)$, $\tilde{u}(g)$ and $v(l)$ with derivatives satisfying: $u' > 0$, $u'' \leq 0$, $\tilde{u}' > 0$, $\tilde{u}'' < 0$ and $v' > 0$, $v'' > 0$ and where the private good consumption is denoted c, the old age public expenditures g and the individual labor supply l. Households have perfect foresights. The wage is denoted w_t and interest factor R_t. The program of each household is:

$$
\begin{cases}
\underset{c_{t+1}; \, l_t}{Max} \; u(c_{t+1}) + \tilde{u}(g_{t+1}) - v(l_t) \\
c_{t+1} = R_{t+1}(1 - \tau)w_t l_t \\
g_{t+1} = \tau w_{t+1} l_{t+1}
\end{cases}
\tag{1}
$$

$$
\Longleftrightarrow \quad \underset{l_t}{Max} \; u(R_{t+1}(1 - \tau)w_t l_t) + \tilde{u}(g_{t+1}) - v(l_t)
\tag{2}
$$

First order condition gives:

$$
R_{t+1}(1 - \tau)w_t u'(c_{t+1}) - v'(l_t) = 0
\tag{3}
$$

Second order condition necessitates:

$$
(R_{t+1}(1 - \tau)w_t)^2 \, u''(C_{t+1}) - v''(l_t) < 0
\tag{4}
$$

Under assumptions on u and v, it's always true.

Technology is Leontiev type $y_t = \min(\alpha l_t, \beta k_t)$ where α and β are respectively capital and labor average productivities. Full capacity utilization implies: $y_t = \alpha l_t = \beta k_t$. Profit is nil. A unit of output is distributed between labor and capital as: $\frac{w_t}{\alpha} + \frac{R_t}{\beta} = 1$. At equilibrium state, the remaining output is stored as productive capital:

$$
k_{t+2} = y_{t+1} - c_{t+1} - g_{t+1} = (1 - \tau)w_{t+1}l_{t+1}
\tag{5}
$$

A little computation gives the following wages and interest factor dynamics:

$$\begin{cases} w_t = \frac{\alpha}{\beta.(1-\tau)}\frac{k_{t+1}}{k_t} \\ R_{t+1} = \beta - \frac{1}{(1-\tau)}\frac{k_{t+2}}{k_{t+1}} \end{cases} \tag{6}$$

Since $y_{t+1} = \beta.k_{t+1}$, one has:

$$\begin{cases} c_{t+1} = \beta.k_{t+1} - \frac{1}{1-\tau}k_{t+2} \\ l_t = \frac{\beta}{\alpha}k_t \end{cases} \tag{7}$$

Then, (3) becomes:

$$\left(\beta.k_{t+1} - \frac{1}{1-\tau}k_{t+2}\right)u'\left(\beta k_{t+1} - \frac{1}{1-\tau}k_{t+2}\right) = \frac{\beta}{\alpha}k_t v'\left(\frac{\beta}{\alpha}k_t\right) \tag{8}$$

Let $U(c) = cu'(c)$ and $V(l) = lv'(l)$, (8) gives:

$$\beta.k_{t+1} - \frac{1}{1-\tau}k_{t+2} = U^{-1}\left(V\left(\frac{\beta}{\alpha}k_t\right)\right) \tag{9}$$

Let $x_t = k_{t-1}$, (9) and balanced budget in each period gives:

$$\begin{cases} k_{t+1} = (1-\tau)\beta k_t - (1-\tau)H\left(\frac{\beta}{\alpha}x_t\right) \\ x_{t+1} = k_t \\ g_{t+1} = \frac{\tau}{1-\tau}k_{t+2} \end{cases} \tag{10}$$

with $H = U^{-1}oV$. The Reichlin's case is $\tau = 0$. In this case, $U(c_{t+1}) = V\left(\frac{\beta}{\alpha}k_t\right)$.

2.2 STEADY STATE EXISTENCE CONDITIONS:

Proposition 2.1:
 Under $\beta > \frac{1}{1-\tau}$ [C1] and $H(l) - lH'(l) > 0$ [C2] or $H(l) - lH'(l) < 0$ [C2'], there is a unique positive solution from (10), $\bar{k} = \frac{\alpha}{\beta}\psi^{-1}\left(\frac{(1-\tau)\frac{\beta}{\alpha}}{\beta(1-\tau)-1}\right)$ with $\psi(l) = \frac{l}{H(l)}$ and $\bar{k} = 0$ is a trivial solution.

Proof
 The fixed points are solutions to $(\beta - \frac{1}{1-\tau})\bar{k} = H\left(\frac{\beta}{\alpha}\bar{k}\right)$. Positive solution exists if and only if $\beta > \frac{1}{1-\tau}$ [C1]. There exists an explicit form for $\bar{k}$ if and only if $\psi(l)$ is monotonic. That is true if $H(l) > lH'(l)$ [C2], then $\psi(l)$ is increasing or if $H(l) < lH'(l)$ [C2'], then $\psi(l)$ is decreasing. ■

Proposition 2.2
 Under conditions [C1] and [C2], the steady state capital increases with τ and under conditions [C1] and [C2'], the steady state capital decreases with τ.

Proof

$\frac{\partial \bar{k}}{\partial \tau} = \frac{1}{(\beta(1-\tau)-1)^2} \psi^{-1\prime} \left(\frac{(1-\tau)\frac{\beta}{\alpha}}{\beta(1-\tau)-1} \right) > 0$ if ψ^{-1} is increasing and $\frac{\partial \bar{k}}{\partial \tau} = \frac{1}{(\beta(1-\tau)-1)^2} \psi^{-1\prime} \left(\frac{(1-\tau)\frac{\beta}{\alpha}}{\beta(1-\tau)-1} \right) < 0$ if ψ^{-1} is decreasing. ∎

3 Permanent business cycle

3.1 CYCLES CONDITIONS

Proposition 3.1
In a neighborhood of the steady state, the dynamics of the economy will be cyclical if and only if:

$$(\alpha, \beta, \tau) \in \left\{ (\alpha, \beta, \tau) \in \mathbb{R}^{*+2} \times [0,1[\, / \, (1-\tau)\beta < 2 \text{ et } H'\left(\frac{\beta}{\alpha}\bar{k}\right) = \frac{\frac{\alpha}{\beta}}{1-\tau} \right\} [C3].$$

Proof
In a neighborhood of $(\bar{k}, \bar{x})$, the dynamics of the system are approximated by:

$$\begin{pmatrix} dk_{t+1} \\ dx_{t+1} \end{pmatrix} \simeq P. \begin{pmatrix} dk_t \\ dx_t \end{pmatrix} \qquad (11)$$

with $dk = k - \bar{k}$, $dx = x - \bar{x}$ and

$$P = \begin{bmatrix} (1-\tau)\beta & -(1-\tau)\frac{\beta}{\alpha}H'(\frac{\beta}{\alpha}\bar{k}) \\ 1 & 0 \end{bmatrix}$$

There are two conditions on eigenvalues for cycles. The first one is that eigenvalues module must be equal to 1, $(1-\tau)\frac{\beta}{\alpha}H'(\frac{\beta}{\alpha}\bar{k}) = 1$ and the second one is that eigenvalues must be complex $(1-\tau)\beta < 2$. ∎

Proposition 3.2
Under $[C1]$, $[C2]$ / $[C2']$ and $[C3]$ the capital cycle frequency increases with respect to τ.

Proof
Under $[C3]$, there exists an orthogonal matrix N such as (11) becomes:

$$\begin{pmatrix} dk_{t+1} \\ dx_{t+1} \end{pmatrix} \simeq N^{-1} \begin{pmatrix} \lambda & 0 \\ 0 & \bar{\lambda} \end{pmatrix} N \begin{pmatrix} dk_t \\ dx_t \end{pmatrix} \qquad (12)$$

where λ and $\bar{\lambda}$ are the eigenvalues of P.

Set θ be the argument of the eigenvalues. One has: $\lambda = \cos\theta + i\sin\theta$ and $\bar{\lambda} = \cos\theta - i\sin\theta$ with $\theta = \arccos\left(\frac{\beta(1-\tau)}{2}\right)$. The cycle period $T = \frac{2\pi}{\theta}$ is decreasing with respect to θ and the frequency θ is increasing with respect to τ: $\frac{\partial\theta}{\partial\tau} = \frac{\frac{\beta}{2}}{\sqrt{1-\left(\frac{\beta(1-\tau)}{2}\right)^2}}$. ∎

3.2 Stability of Reichlin type dynamics

Let f_a be a map $f_a : \begin{pmatrix} k \\ x \end{pmatrix} \mapsto \begin{pmatrix} ak - h_a(x) \\ k \end{pmatrix}$ defining a Reichlin type dynamics: $(k_{t+1}, x_{t+1}) = f_a(k_t, x_t)$. There exists a neighborhood of a' where $|\lambda(a')| = 1$ such that f_a is a diffeomorphism mapping an open set of $\mathbb{R}^{+2}$ onto an open set of $\mathbb{R}^{+2}$. Under conditions $[C1], [C2]/[C2'], [C3]$, let $\lambda(a)$ and $\overline{\lambda(a)}$ the two non-real eigenvalues of $P = \begin{bmatrix} a & h'_a \\ 1 & 0 \end{bmatrix}$. There exists a new basis for $dk = k - \bar{k}$ and $dx = x - \bar{x}$ where the system can be written as:

$$\begin{pmatrix} \hat{k}_{t+1} \\ \hat{x}_{t+1} \end{pmatrix} = \begin{pmatrix} \operatorname{Re}\lambda(a') & \operatorname{Im}\lambda(a') \\ -\operatorname{Im}\lambda(a') & \operatorname{Re}\lambda(a') \end{pmatrix} \begin{pmatrix} \hat{k}_t \\ \hat{x}_t \end{pmatrix} - \begin{pmatrix} \zeta_a(x_t) \\ 0 \end{pmatrix} \quad (13)$$

$$\text{with} \quad \begin{cases} \begin{pmatrix} \hat{k}_t \\ \hat{x}_t \end{pmatrix} = \begin{pmatrix} \frac{1}{\sin\theta} & -\frac{\cos\theta}{\sin\theta} \\ 0 & 1 \end{pmatrix} \begin{pmatrix} dk_t \\ dx_t \end{pmatrix} \\[2ex] \begin{pmatrix} \zeta_a(x_t) \\ 0 \end{pmatrix} = \begin{pmatrix} \frac{1}{\sin\theta} & -\frac{\cos\theta}{\sin\theta} \\ 0 & 1 \end{pmatrix} \begin{pmatrix} \varphi_a(x_t) \\ 0 \end{pmatrix} \end{cases}$$

where $\varphi_a(x_t) = \frac{1}{2}h''_a(\bar{x})(x_t - \bar{x})^2 - \frac{1}{6}h'''_a(\bar{x})(x_t - \bar{x})^3 + O\left((x_t - \bar{x})^5\right)$.

Let Φ_a be the map defined by (13), Φ_a is one parameter family of diffeomorphism on $\mathbb{R}^2$ satisfying the following conditions for a in near a':
(i) $\Phi_a(0,0) = (0,0)$
(ii) $d\Phi_a(0,0)$ has two non real eigenvalues $\lambda(a), \overline{\lambda(a)}$ with $|\lambda(a')| = 1$
(iii) $\frac{d|\lambda(a)|}{da} > 0$
(iv) $(\lambda(a'))^m \neq 1$ for $m = 1,2,3,4,5$.

Write $\Phi_{a'} = \Phi$ and $\lambda(a') = \lambda$. By setting $z = \hat{k} + i\hat{x}$, Φ can be written as:

$$\begin{aligned} \Phi(z) &= \left(\lambda z + \frac{G_{11}}{2}z^2 + G_{12}z\bar{z} + \frac{G_{22}}{2}\bar{z}^2\right) \\ &\quad + \left(\frac{G_{111}}{6}z^3 + \frac{G_{112}}{2}z^2\bar{z} + \frac{G_{122}}{2}z\bar{z}^2 + \frac{G_{111}}{2}\bar{z}^3\right) + O\left(|z|^5\right) \end{aligned}$$

More precisely, for Reichlin type dynamics, we have:

$$\Phi(z) = (\cos\theta + i\sin\theta)\, z - \frac{\cos\theta}{\sin\theta}\left[\tfrac{1}{2}h_a''(\bar{x})\left(\tfrac{z-\bar{z}}{2i}\right)^2 - \tfrac{1}{6}h_a'''(\bar{x})\left(\tfrac{z-\bar{z}}{2i}\right)^3 + O\left(|z^4|\right)\right]$$

Theorem 3.3 (Lanford-Wan)

$\Phi(z) = N\Phi(z) + O\left(|z|^5\right)$, where in polar coordinates, $N\Phi : (r,\phi) \mapsto$
$\left((1+\tilde{a})\,r - f_1(a)\,r^3, \phi + \theta(a) + f_3(a)\,r^2\right)$ with $(1+\tilde{a}) = |\lambda(a)|$. Furthermore, $f_1(a)$ is given by:

$$\mathrm{Re}\left[\frac{(1-2\lambda)\bar{\lambda}}{2(1-\lambda)}G_{11}G_{12}\right] + \frac{1}{2}G_{12}\bar{G}_{12} + \frac{1}{4}G_{22}\bar{G}_{22} - \mathrm{Re}\left(\frac{\bar{\lambda}G_{112}}{2}\right)$$

Theorem 3.4 (Ruelle-Takens)

If $f_1(a') > 0$, then for all sufficiently small positive $\tilde{a}$, Φ_a has an attracting invariant circle.

Corollary 3.5

In the case of a dynamic system (10) the condition for an invariant attractive circle is $(H_a'')^2 - H_a''' > 0$ $[C4]$, in a neighborhood of a', with $a' = (1-\tau)\beta$.

3.3 PERMANENT BUSINESS CYCLE CONDITION

When τ changes, the $[C3]$ cycle condition cannot stay verified. That's why, we propose to endogenize α with respect to τ and β.

Proposition 3.6

Under $[C1]$, $[C2]$ / $[C2']$ and $y\psi^{-1\prime}(y)\,H''\left(\psi^{-1}(y)\right) > -H'\left(\psi^{-1}(y)\right)$
$[C5]$ or $y\psi^{-1\prime}(y)\,H''\left(\psi^{-1}(y)\right) > -H'\left(\psi^{-1}(y)\right)$ $[C5']$ where $y = \frac{(1-\tau)\frac{\beta}{\alpha}}{\beta(1-\tau)-1}$,
if $\alpha = (1-\tau)\beta\Gamma(s)$ with $\Gamma(s) = \frac{s}{\Theta^{-1}(s)}$, $s = \frac{1}{\beta(1-\tau)-1}$ and $\Theta(y) = yH'\left(\psi^{-1}(y)\right)$ then $[C3]$ is always true.

Proof:
$\Theta(y)$ must be inversive. If $\Theta'(y) > 0$ or if $\Theta'(y) < 0$, that's true. ∎

How to justify this relation? A simple way is to suppose that it is the result of a very special maximization behavior. That's why, we add these new assumptions:
i) there exists a relation between labor productivity and collective effort ;
ii) workers prefer to have or not to have a pay-as-you-go system ;

iii) workers choose effort level and labor supply independently. They know that minimal effort is not optimal because there is a collective but not individual positive link between wages and effort.

Proposition 3.7

Under $[C1]$, $[C2]$ / $[C2']$, $[C3]$ *and* $[C5]$ / $[C5']$ *there always exists a functional form of utility* $Z(e,\beta,\tau) = -\frac{1}{2}\alpha(e)^2 + (1-\tau)\beta\Gamma\left(\frac{1}{\beta(1-\tau)-1}\right)$ $\alpha(e)$ *where* $\alpha(e)$ *is the relation between effort and labor productivity with* $\alpha'(e) > 0$ *and* $\alpha''(e) \leq 0$ *such as the optimal condition* $\frac{\partial Z(e,\beta,\tau)}{\partial e} = 0$ *implies cycle.*

Proof:
One can verify easily that $\frac{\partial Z(e,\beta,\tau)}{\partial e} = 0$ gives cycle condition $\alpha = (1-\tau)\beta\,\Gamma(s)$. Moreover, this solution is optimal because $\frac{\partial^2 Z(e,\beta,\tau)}{\partial e^2} = -\alpha'^2(e) < 0$. ∎

This condition is restrictive but it is necessary for our purpose. With this additional utility form, the economic behavior of households offsets enough changes of τ by raising or decreasing their effort to maintain cycle condition. From now on, as $\alpha = (1-\tau)\beta\Gamma(s)$, under $[C1]$, $[C2]$ / $[C2']$, $[C3]$, $[C4]$ and $[C5]$ / $[C5']$, steady capital level change with respect to τ is given by :

$$
\begin{aligned}
\frac{d\bar{k}}{d\tau} &= \frac{\partial\bar{k}}{\partial\tau} + \frac{\partial\bar{k}}{\partial\alpha}\frac{\partial\alpha}{\partial\tau} \\
&= \frac{1}{(\beta(1-\tau)-1)^2}\psi^{-1\prime}\left(\frac{(1-\tau)\frac{\beta}{\alpha}}{\beta(1-\tau)-1}\right) - \frac{(1-\tau)\frac{\beta}{\alpha^2}}{(\beta(1-\tau)-1)}\psi^{-1\prime}\left(\frac{(1-\tau)\frac{\beta}{\alpha}}{\beta(1-\tau)-1}\right) \\
&\quad \times \left(-\beta\Gamma(s) + \frac{(1-\tau)\beta}{(\beta(1-\tau)-1)^2}\Gamma'(s)\right) \\
&= \frac{1}{(\beta(1-\tau)-1)^2}\left(1 + (1-\tau)\frac{\beta}{\alpha^2}(\beta(1-\tau)-1)\right) \\
&\quad \left(\beta\Gamma(s) - \frac{(1-\tau)\beta}{(\beta(1-\tau)-1)^2}\Gamma'(s)\right) \times \psi^{-1\prime}\left(\frac{(1-\tau)\frac{\beta}{\alpha}}{\beta(1-\tau)-1}\right)
\end{aligned}
$$

4 Example

4.1 THE MODEL

Suppose $v(l) = \frac{a}{2}l^2 + bl$; $u(c) = c$ and[3] $\tilde{u}(g) = \lambda g$. The household economic behavior is the solution to $\underset{e,c_{t+1},l_t}{Max}\ u(c_{t+1}) + \tilde{u}(g_{t+1}) - v(l_t) + $

---

[3]If $\lambda = 1$, public good and private good are perfectly interchangeable and the pay-as-you-go system is equivalent to an traditional no contributive unfunded pension system. So τ can be considerd like an intergenerational transfert rate.

$\gamma Z(\underset{+,-}{e},\tau)$ such that $c_{t+1} = R_{t+1}(1-\tau)w_t l_t$, $g_{t+1} = \tau w_{t+1} l_{t+1}$ with

$Z(\underset{+,-}{e},\tau) = -\dfrac{e^2}{2} + \dfrac{2(a_2\tau + b_2)}{(a_1\tau + b_1)}e$ and $\alpha(e) = \delta e$, the relationship between effort and productivity. The second order condition are always verified because $u''(c) - v''(l) = -a < 0$ and $z''(e) = -1 < 0$.

Hence, by using capital per head dynamic equation (10), we can write:

$$\begin{cases} k_{t+2} = (1-\tau)\beta k_{t+1} - (1-\tau)\frac{\beta}{\alpha}x_{t+1}^2 \\ x_{t+2} = k_{t+1} \end{cases}$$

and $e(\tau) = \hat{e} = \dfrac{a_2\tau + b_2}{2(a_1\tau + b_1)}$. There exists two steady states A et B:

$$A = \{\bar{k};\bar{x}\} = \{0;0\} \text{ et } B = \{\bar{k};\bar{x}\} = \left\{ \dfrac{(1-\tau)[\beta - b(\frac{\beta}{\alpha})] - 1}{a(1-\tau)\left(\frac{\beta}{\alpha}\right)^2} ; \dfrac{(1-\tau)[\beta - b(\frac{\beta}{\alpha})] - 1}{a(1-\tau)\left(\frac{\beta}{\alpha}\right)^2} \right\}$$

In a neighborhood of the steady state B, the Jacobian matrix of this system is:

$$J = \begin{bmatrix} (1-\tau)\beta & -\frac{\beta}{\alpha}(1-\tau)\left[2a\bar{x} + b\right] \\ 1 & 0 \end{bmatrix}$$

$$|J| = \frac{\beta}{\alpha}(1-\tau)\left[2a\bar{x} + b\right] = (1-\tau)\beta\left(2 - \frac{b}{\alpha}\right) - 2$$

The dynamics of the economy will be cyclical if and only if

- $\alpha = \dfrac{2\beta b(1+\tau)}{2\beta(1-\tau) - 3}$. This relation allows to obtain eigenvalues module equal to 1 in absolute value. If $\alpha = \delta e$, permanent cycle condition necessitates $a_1 = -\beta$; $b_1 = \dfrac{2\beta - 3}{2}$; $a_2 = b_2 = \dfrac{\beta b}{\delta}$.

- $\dfrac{3}{2(1-\tau)} < \beta < \dfrac{2}{(1-\tau)}$. This last condition allows eigenvalues to be complex and $\alpha > 0$. However, there is not β verifying this condition $\forall \tau$: $\dfrac{1,5(b+\tau)/(1-\tau)}{2\beta(1-\tau) - 3} < \alpha < \dfrac{2(b+\tau)/(1-\tau)}{2\beta(1-\tau) - 3}$. In this case, the cycle stability is always verified because: $\dfrac{(4 - (1-\tau)\beta)}{\left(4 + ((1-\tau)\beta)^2\right)} > 0$.

4.2 COMPUTING SIMULATIONS

Figures 1, 2 and 3 illustrate these purposes for the following parameters values:

Utility parameters	labor disutility $\begin{cases} a = 0.003 \\ b = 0.1 \end{cases}$
	public good utility $\lambda = 1$
	effort utility $\gamma = 100$
Production parameters	capital productivity $\beta = 2.1$
	labor productivity and effort $\delta = 1$
Initial values	$k_0 = k_1 = 1000$

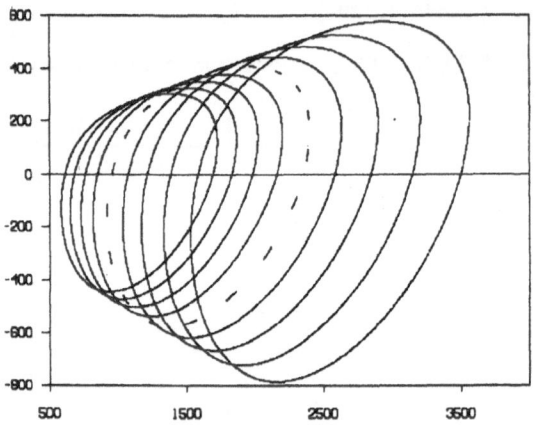

Figure 1 - Capital per head phases diagram for $\tau = 0.26, ..., 0.268$

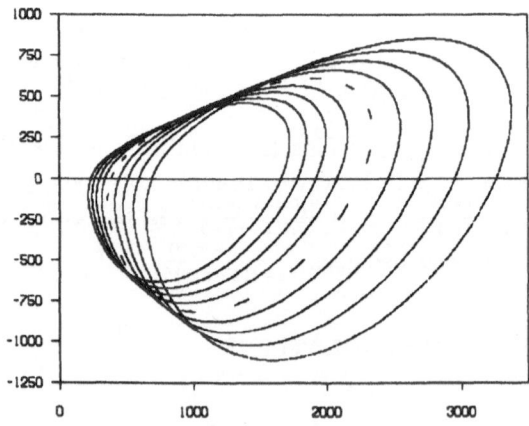

Figure 2 - Private Good Consumption per head phases diagram for
$\tau = 0.26, ..., 0.268$

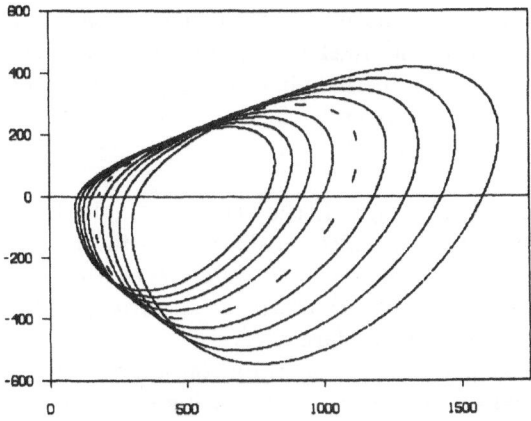

Figure 3 - Life Cycle Welfare phases diagram for $\tau = 0.26, ..., 0.268$

Figure 3 gives welfare phases diagram. The elliptic forms moving to the right characterize the τ rising. That's also true for figures 1 and 2. Two sorts of conflict are interesting: a conflict between efficiency (variation of the steady welfare state level which is identified by the center of a elliptic form) and equity (variations of the welfare range) criteria and a conflict between two equity criteria (variations of the welfare range and variations of the lowest welfare level). Indeed, when τ rises, the graphs on figure 3 show that the steady welfare state level, the lowest welfare level and the welfare range rise. Therefore, the Rawlsian criterium consisting in maximizing the lowest welfare level is increasing with respect to τ. However, the egalitarian criterium consisting in reducing the intergenerational welfare range is decreasing with respect to τ. Finally, the efficiency criterium consisting in maximizing the steady welfare state level is increasing with respect to τ.

Time paths are not perfectly periodic because this model is not linear and chaotic time paths exist. It is why on the graphs, we can see that all the values of the interval $[k_{\min}; k_{\max}]$ seem to be on the path. Then, discrete paths seem to look like to continuous time cycles.

5 Conclusion

In this article, we obtain a permanent business cycle by endogenizing the labor productivity with respect to payroll tax rate to conserve the business cycle. The permanent endogenous cycle allows us to contribute to

122

the debate about equity and efficiency of the pay-as-you-go system when the payroll tax rate is modified in an original fashion. Indeed, by using computing simulation, an increase in the intergenerational transfer rate i.e. the payroll tax rate can have two effects on welfare:

i) first on steady welfare state i.e. the efficiency reference level which rises;

ii) second on welfare range. If the range rises, the increase of τ is not improving according to the pure egalitarian criterium. On the contrary, if the increase of one half range is inferior to the increase of steady welfare state, the increase of τ is improving according to Rawlsian criterium because the lowest level of welfare rises.

One of our next perspectives of study consists in specifying analytically these latter found effects by using computing simulations. Notably, it can be done by writing economic variables time paths (capital, consumption and welfare) in the following way: $x_t = \bar{x} + A \cos(\theta t + \varphi)$ where the range is denoted A, the frequency θ, the phase φ and the capital, consumption or welfare steady state $\bar{x}$.

Another perspective consists in suppressing the permanent economic cycle condition and in investigating all chaotic time paths with respect to τ.

References

[1] ALLAIS M. [1947], *Economie et intérêt*, Imprimerie Nationale, Paris.

[2] BARRO R.J. [1974], "Are Gouvernement Bonds Net Wealth?", *Journal of Political Economy*, 82, 1095-1117.

[3] BENHABIB J. and R. DAY [1982], "A Characterization of Erratic Dynamics in the Overlapping Generations Model", *Journal of Economic Dynamics and Control*, 4, 37-55.

[4] BREYER F. and M. STRAUB [1993], "Welfare Effects of Unfunded Pension Systems when Labor Supply is Endogenous", *Journal of Public Economics*, 50 , 77-91.

[5] DIAMOND P. A. [1965], "National Debt in a Neoclassical Growth Model", *American Economic Review*, 55 , 1126-1150.

[6] GALE D. [1973], "Pure Exchange Equilibrium of Dynamic Economic Models", *Journal of Economic Theory*, 6 , 2-36.

[7] GALOR O [1992], "A Two-Sector Overlapping Generations Model: A Global Characterization of the Dynamical System", *Econometrica*, 60, 1351-1386.

[8] GRANDMONT J.M. [1985], "On Endogenous Competitive Business Cycles", *Econometrica*, 53, 995-1045.

[9] LANFORD O.E. [1972], "Bifurcation of Periodic Solutions into Invariant Tori: the Work of Ruelle and Tackens", in *Nonlinear Problems in the Physical Sciences and Biology*, Lectures Notes in Mathematics, n°322, Springer-Verlag, 159-192.

[10] PHELPS E.S. and J. C. RILEY [1978], "Rawlsian Growth: Dynamic Programming of Capital and Wealth for Intergenerational "Maximin" Justice", *Review of Economic Studies*, 45, 103-120.

[11] RAWLS J. [1971], *A Theory of Justice*, Harvard University Press, Cambridge.

[12] REICHLIN P. [1986], "Equilibrium Cycles in an Overlapping Generations Economy with Production", *Journal of Economic Theory*, 40, 89-102.

[13] RODRIGUEZ A. [1981], "Rawls's Maximin Criterion and Time Consistency", *Review of Economic Studies*, 48, 599-605.

[14] RUELLE D. and F. TAKENS [1971], "On the Nature of Turbulence", *Communications in Mathematical Physics*, 20, 167-192.

[15] SAMUELSON P.A. [1958], "An Exact Consumption-Loan Model of Interest with or without the Social Contrivance of Money", *Journal of Political Economy*, 66, 467-482.

[16] WAN Y.H. [1978], "Computation of the Stability Condition for the Hopf Bifurcation of Difeomorphisms on $\mathbb{R}^2$", *SIAM Journal of Applied Mathematics.*, 34 , 167-175.

[17] ZHANG J. [1995], "Social Security and Endogenous Growth", *Journal of Publics Economics*, 58 , 185-213.

Part 3

Keynesian Models

Relaxation Cycle, Chaotic Dynamics and Limit Cycle: a Model with Keynesian "Flavour"

Gilbert ABRAHAM-FROIS and Edmond BERREBI

1 Introduction

R. Goodwin (1951) was probably the first to introduce "relaxation cycles" in macroeconomics; Hicks' analysis is contemporary and indeed previous (1950), both authors introducing "non-linear accelerator" in their analysis; but Goodwin' originality is obvious since he has used relaxation cycle and Liénard-van der Pol equations (as soon as 1933, Ph. Le Corbeiller had insisted on their interest and relevance). Further work has shown that in many cases relaxation cycle could be considered as a stable "attractor", a rather peculiar "limit-cycle", named in some cases (Chiarella 1990) "limiting limit cycle".

In other respects, one knows that cubic function can give rise to chaotic dynamics; in previous work - Puu (1991) and extension by Abraham-Frois and Berrebi (1995) - it appears possibility of having - simultaneously !! - chaotic dynamics and relaxation cycle. Moreover, in Day and Schafer (1986), IS-LM model can produce chaotic dynamics. The aim of this paper is to give further extension to this kind of analysis. Some interesting results appear: first, the relationship between cycle and chaotic dynamics is more complex than expected; second, the relaxation cycle is also an attractor, a limit cycle; third, this particular cycle is locally stable, so that one can contemplate some connection between exogeneous shocks and endogeneous cycles.

2 A macro-model with keynesian flavour

Puu's model is a "keynesian standard" type model and may be situated as an extension of previous analysis by Samuelson (1939), Hicks (1950) and Goodwin (1951). Consumption and investment function are rather specific.

More precisely, Puu assumes that investment at time t, I_t is a *non-linear* function of finite differences of incomes at $(t-1)$ and can be written as:

$$I_t = v(Y_{t-1} - Y_{t-2}) - v(Y_{t-1} - Y_{t-2})^3$$

The graph has the following characteristics: it goes through the origin with slope v; this slope decrease slowly till stabilization; it turns over after for higher differences of incomes, which does not seem quite relevant. Moreover, in case of decreases of income, this function gives rise to negative investment, which is at least annoying.

To avoid these problems, we suggest to introduce an autonomous investment so that the investment function now writes:

$$I_t = v(Y_{t-1} - Y_{t-2}) - v(Y_{t-1} - Y_{t-2})^3 + I_0 \qquad \text{(1a)}$$

There is no loss of generality by writting $I_0 = 1$, since this means a change of unit in income level; hence the investment function suggested :

$$I_t = v(Y_{t-1} - Y_{t-2}) - v(Y_{t-1} - Y_{t-2})^3 + 1 \qquad \text{(1b)}$$

By writting $Z_t = \Delta Y_t = Y_t - Y_{t-1}$, one obtains:

$$I_t = vZ_{t-1} - vZ_{t-1}^3 + 1 \qquad \text{(2)}$$

Investment function is a cubic, which is essential for following analysis.

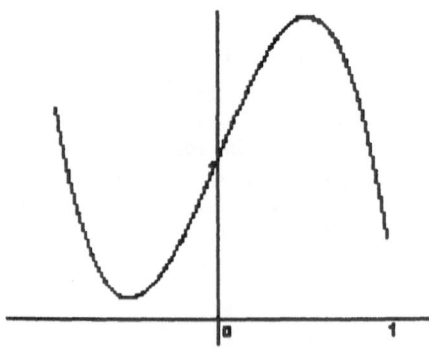

Figure 1: Investment with cubic function and autonomous investment

A second assumption, quite central here, concerns the consumption - or saving - function. s being the propensity to save, one makes the assumption that consumption in period t depends upon two elements: the first one is the level of income of $(t-1)$, the second being the amont of saving in $(t-2)$. Part ε of saving sY_{t-2} is consumed in t; part $(1-\varepsilon)sY_{t-2}$ is never consumed, so that one can define $(1-\varepsilon)s$ as the "rate of eternal saving" (the exact value of this parameter is not very important; what really matters is to know wheter ε is equal to or different from one).

So, consumption at t is the sum of part of income of $t-1$, $(1-s)Y_{t-1}$ and of part of saving appeared in $(t-2)$, εsY_{t-2} hence

$$C_t = (1-s)Y_{t-1} + \varepsilon sY_{t-2} \tag{3}$$

or

$$C_t = Y_{t-1} - s(Y_{t-1} - Y_{t-2}) - (1-\varepsilon)sY_{t-2} \tag{4}$$

or

$$C_t = Y_{t-1} - sZ_{t-1} - (1-\varepsilon)sY_{t-2} \tag{5}$$

From the equilibrium relation,

$$Y_t = C_t + I_t \tag{6}$$

on gets, by replacing C_t and I_t by their value in 5 or 3 :

$$
\begin{aligned}
Y_t &= C_t + I_t \tag{7}\\
&= Y_{t-1} - sZ_t - (1-\varepsilon)sY_{t-2} + vZ_{t-1} - vZ_{t-1}^3 + 1 \tag{8}\\
&\Leftrightarrow\\
Y_t - Y_{t-1} &= -sZ_t - (1-\varepsilon)sY_{t-2} + vZ_{t-1} - vZ_{t-1}^3 + 1\\
&\Leftrightarrow\\
Z_t &= -sZ_t - (1-\varepsilon)sY_{t-2} + vZ_{t-1} - vZ_{t-1}^3 + 1
\end{aligned}
$$

One can write $q = v - s$, difference between accelerator v and propensity to consume s. So this equation can now be written as :

$$Z_t = qZ_{t-1} - (q+s)Z_{t-1}^3 - (1-\varepsilon)sY_{t-2} + 1 \tag{9}$$

We have now to distinguish the case where rate or eternal savings is equal to zero and from cases when it is different from zero becomes smaller and smaller.

3 When "eternal rate of saving" is equal to zero: $(1 - \varepsilon)\, s = 0$

In this case, on which we will skip rather quickly, one finds a kind of chaotic dynamics which is rather well known; main characteristics will be shown as follows. Relation (8) becomes:

$$Z_t = qZ_{t-1} - (q + s)\, Z_{t-1}^3 + 1 \tag{8'}$$

which is a first order difference equation in Z. In this model, sY_{t-2} appeared in period $t - 2$ is *totally consumed* in t : as a matter of fact, in each period, one saves s per cent of current income which is left untouched next period and entirely consumed two periods later. The evolution of the general term Z_t of the recurrent series (Z_t) can be studied from the properties of f function and of its iteratives $f^2, ..., f^n, ...$, i.e. orbits of function f defined by :

$$z \to f(z) = qz - (q + s)\, z^3 + 1 \tag{10}$$

Particular points of function f graph
f function takes for

- $z = 0$ value $f(0) = 0 \cdot q - 0 \cdot (q + s) + 1 = 1$

- $z = 1$ value $f(1) = 1 \cdot q - 1 \cdot (q + s) + 1 = 1 - s$

- $z = -1$ value $f(-1) = -1 \cdot q - (-1) \cdot (q + s) + 1 = 1 + s$

So graphs goes through points $(-1, 1 + s), (0, 1)$ and $(1, 1 - s)$.
Critical points of f function
Derivative $f'(z) = q - 3(q + s)\, z^2$ has two roots,
$z_1^* = \sqrt{\dfrac{q}{3(q + s)}}$ et $z_2^* = -\sqrt{\dfrac{q}{3(q + s)}}$ which are the two critical points
of function f. Since, $f(z_1^*) = \left[q - (q + s)\,(z_1^*)^2\right] z_1^* + 1 = \dfrac{2q}{3\sqrt{3}}\sqrt{\dfrac{q}{q + s}} + 1$
and $f''(z_1^*) = -6(q + s)\, z_1^* < 0$, point $(z_1^*, f(z_1^*))$ is a *local maximum*;
in the same way, $f(z_1^*) = f(-z_2^*) = -\left[q - (q + s)\,(z_1^*)^2\right] z_1^* + 1 = -\dfrac{2q}{3\sqrt{3}}\sqrt{\dfrac{q}{q + s}} + 1$ and since $f''(z_2^*) = 6(q + s)\, z_1^* > 0$, point $(z_2^*, f(z_2^*))$ is
a *local minimum*.
All terms of the series $\{z, f(z), f^2(z), ..., f^n(z), ...\}$ where $z > 0$ will be positive if local maximum $f(z_1^*) < Z_1$ where Z_1 is the positive real root of $f(z) = qz - (q + s)\, z^3 + 1 = 0$.

Special case $I_0 = 0$

When in equation (1a) one choses $I_0 = 0$, it can be shown that for $q < 2.59809$, any orbit generated by f, $\{z, f(z), f^2(z), ..., f^n(z), ...\}$ has all its terms positive if z belongs to interval $[0, Z_1]$ and all its terms negative if z belongs to interval $[-Z_1, 0]$. On the contrary when $q > 2,59810$ mapping $\{z, f(z), f^2(z), ..., f^n(z), ...\}$ has positive and negative terms since some points in the interval $[0, Z_1]$ which are such that $f(z_1) > Z_1$ and consequently $f^2(z) = f[f(z)] < 0$.

In this first case, this means that, according to the initial value of z (which means - let us remind ...- a change in income level), one has either monotonous increase, either monotonous decrease in the level of income. For $q = 2,59809...$ it appears a bifurcation in the system with alternance of periods of expansion and recession.

We will return, later on, on the importance of this distinction. It may be useful to draw the "bifurcation diagram".

Bifurcation diagram for function f with $I_0 = 0$

Let us plot the bifurcation diagram for f function; on the $x-$absciss, we report the values of parameter q which can vary between 0 to 3 and along the $y-$absciss the 350 points of $\{z, f(z), f^2(z), ..., f^n(z), ...\}$ for function f associated to the value of different values of q end for values of n going from 150 to 1500.

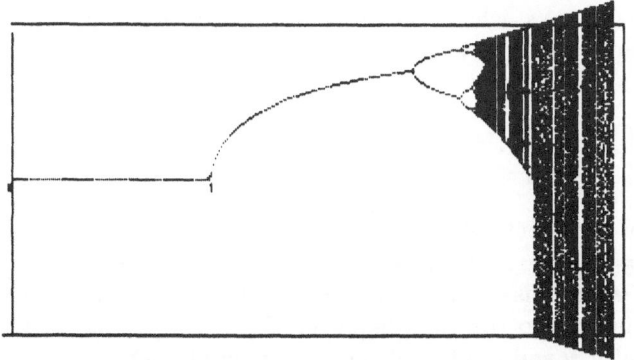

Figure 2: Bifurcation diagram for $q \in [0, 3]$

Nota : when $I_0 = 0$, the model's dynamics depends only on the value of $q = v - s$. On the other hand, when $I_0 = 1$, stability of q does not exclude a change in the system's dynamics when s takes differents values.

4 When "eternal rate of savings" is not equal to zero: $(1 - \varepsilon)\, s \neq 0$ without however being insignificant ...

The assumption of a non-zero eternal rate of savings modifies deeply the dynamic of the system, the difference equation now writes:

$$Z_t = [q + (1 - \varepsilon)\, s]\, Z_{t-1} - (q + s)\, Z_{t-1}^3 - (1 - \varepsilon)\, sY_{t-1} + 1 \qquad (11)$$

The economic interpretation is that is such case saving sY_{-2} of $(t - 2)$ is now divided in two parts:

- εsY_{t-2} which is consumed in period t

- $(1 - \varepsilon)\, sY_{t-2}$ which is saved until the end of the program.

This entails deep changes in the dynamic evolution; there will appear cycles of different kind, which may be associated - or not - with chaotic evolution; moreover, another difference appears: in the first model, the general evolution depended only on $q = v - s$; now, it depends first on v et second on $(1 - \varepsilon)\, s$, which obviously means changes in q.

From equations (2) and (11), one gets the following system of two equations :

$$(S) \quad \begin{cases} Y_t = Y_{t-1} + Z_t \\ Z_t = [q + (1 - \varepsilon)\, s]\, Z_{t-1} - (q + s)\, Z_{t-1}^3 - (1 - \varepsilon)\, sY_{t-1} + 1 \end{cases}$$

One can represent solution of (S) system in the phase diagram (Y_t, Z_t) where $Z_t = Y_t - Y_{t-1}$. The phase diagram is here of the discrete type - using difference equations - when usual presentations uses differential equations. We shall figure too the time-sequence $(Y_1, Y_2, ..., Y_t)$ for particular values of accelerator v and to eternal rate of saving $(1 - \varepsilon)\, s$.

In previous work[1], we have studied the case where v took different values for $(1 - \varepsilon)\, s$ constant; we shall insist specially on the situation where $(1 - \varepsilon)\, s$ tends to zero.

For a special value of v, say $v = 1.75$, one can depict the phase diagram (Y_t, Z_t) and the time sequence Y_t corresponding to different values of eternal rate of saving $(1 - \varepsilon)\, s$.

According to the values of $(1 - \varepsilon)\, s$, it appears for the phase diagram of system (S) either:

---

[1]Cf. Abraham-Frois and Berrebi [1995].

- a finite number of points

- a thin continuous curve

- a thick continuous curve

- a laminated surface

So, it appears, for

$(1 - \varepsilon)\,s$	Forms	Figure
0.08	closed curve	3a
0.25	butterfly	3b
0.45	clouds	3c
0.60	signature	3d
0.70	bracelet	3e
0.85	separate segments	3f

Those phase diagrams presented on fig. 3 are all strange attractors since, when t goes to infinity, points (Y_t, Z_t) go back to these attractors without regularity.

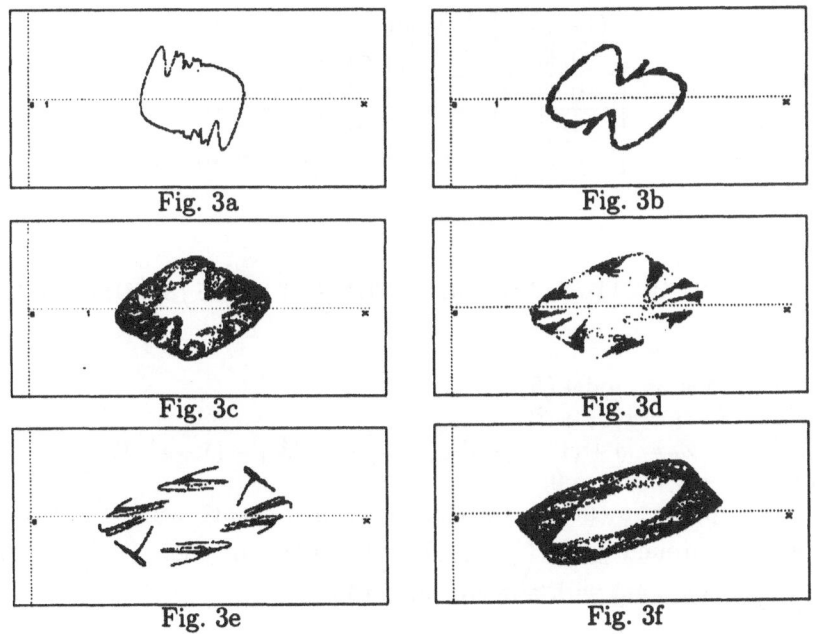

Fig. 3a   Fig. 3b

Fig. 3c   Fig. 3d

Fig. 3e   Fig. 3f

134

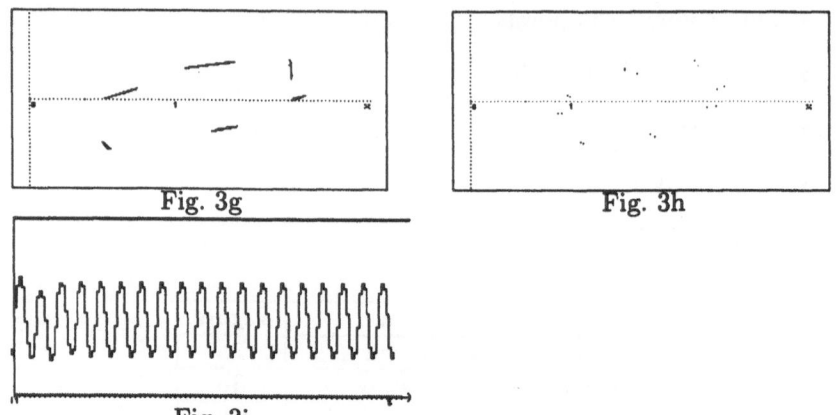

Fig. 3g

Fig. 3h

Fig. 3i

Figure 3: Phase diagram for $v = 1.75$ and $(1 - \varepsilon)\,s$ variable

An another hand, for numerous values of $(1 - \varepsilon)\,s$, one has a finite number of points on the phase diagram which means that one obtains a periodic time sequence (Y_t).

For instance, it comes for :

$(1 - \varepsilon)\,s$	Periodic time sequence
0.20	period 18
0.35	period 12
0.41	period 10
0.55	period 8 (fig. 3h and 3i)

5   When eternal rate is trifling: $(1 - \varepsilon)s$ turns to 0

Let us look at model (S)

$(S) \quad \begin{cases} Y_t = Y_{t-1} + Z_t \\ Z_t = [q + (1 - \varepsilon)s]Z_{t-1} - (q + s)Z_{t-1}^3 - (1 - \varepsilon)sY_{t-1} + 1 \end{cases}$

when $(1 - \varepsilon)s$ turns to 0.

There appears a cubic with bifurcations and chaotic zone. This cubic is symetrical around point $\left(\frac{1}{(1-\varepsilon)s}, 0\right)$ since image of point $\left(\frac{2}{(1-\varepsilon)s} - Y_{t-1}\right.$, $-Z_{t-1})$ is a point with abscissa and ordinate :

$$- [q + (1 - \varepsilon)\,s]\,Z_{t-1} + (q + s)\,Z_{t-1}^3 - (1 - \varepsilon)\,s\left[\frac{2}{(1-\varepsilon)s} - Y_{t-1}\right] + 1$$
$$= - \left[(q + (1 - \varepsilon)s)\,Z_{t-1} - (q + s)\,Z_{t-1}^3 - (1 - \varepsilon)sY_{t-1} + 1\right] = -Z_t$$

So part part corresponding to $Z_t < 0$ (which lies under the OY_t axis) is symetrical of part corresponding to $Z_t > 0$ (which lies above the OY_t axis) around point $\left(\frac{1}{(1-\varepsilon)s}, 0\right)$. Consequently, one shall look at the superior part of the cubic and afterwards find its symmetry around point $\left(\frac{1}{(1-\varepsilon)s}, 0\right)$.

5.1 NEW TYPES OF CYCLES APPEAR NOW: RELAXATION CYCLE

When $(1-\varepsilon)s$ turns to 0 and for instance has for value 0.0001, there appear "relaxation cycles" with or without chaotic dynamic. When starting from point $\left(Y_0 = \frac{1}{(1-\varepsilon)s} = 10000, Z_0 = 0.5\right)$ different cases come to evidence :

- If $v = 1.5$ system (S) gives rise in phase diagram (Y_t, Z_t) to a relaxation cycle limited by straight lines $Y = 8880$ and $Y = 11120$.

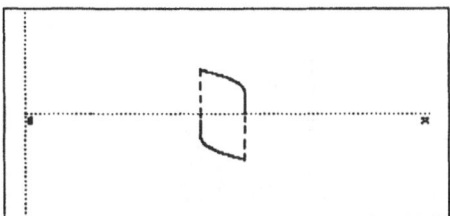

Figure 4: Relaxation cycles for $v = 1.5$

- If $v = 1.75$, there appears a relaxation cycle and a more interesting dynamics with appearance of a double bifurcation system. Cycle is then limited by straight lines $Y = 8100$ and $Y = 11900$.

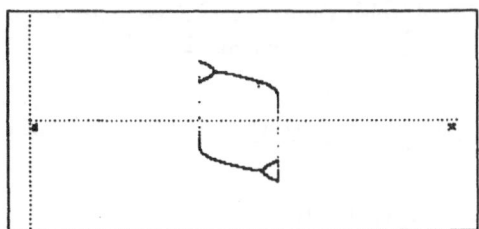

Figure 5: Relaxation cycles and straight lines for $v = 1,75$

- If $v = 1.87$ and always $(1 - \varepsilon)s = 0.0001$, relaxation cycle still exists with many more bifurcations. This cycle is limited by straight lines $Y = 7700$ and $Y = 12300$.

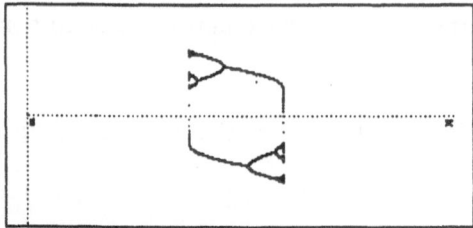

Figure 6: Relaxation cycles for $v = 1.87$

- A last, for $v = 2.12$ and still $(1 - \varepsilon)s = 0.0001$, we have again relaxation cycle and much richer dynamics: there appears double bifurcation processes with extension to dynamic zones. There are two limits, straight lines $Y = 6855$ and $Y = 13145$.

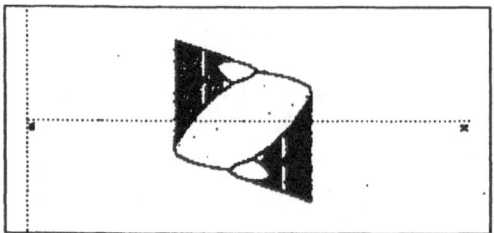

Figure 7: Relaxation cycles for $v = 2, 12$

5.2 Why relaxation cycles?

The appearance of relaxation cycle is easy to explain: in the phase diagram in (Y_t, Z_t), $\frac{\Delta Z_t}{\Delta Y_t} \to \infty$ when $(1 - \varepsilon)s \to 0$. As matter of fact, one has:

$$\frac{Z_t - Z_{t-1}}{Y_t - Y_{t-1}} = \frac{Z_t - Z_{t-1}}{Z_t} = \frac{[q+(1-\varepsilon)s-1]Z_{t-1}-(q+s)Z_{t-1}^3-(1-\varepsilon)sY_{t-1}+1}{Z_t}$$

and with $X_t = (1 - \varepsilon)sY_t$ one finds:

$$\frac{Z_t - Z_{t-1}}{X_t - X_{t-1}} = \frac{Z_t - Z_{t-1}}{(1 - \varepsilon)sZ_t} = \frac{1}{(1 - \varepsilon)s} \frac{[q+(1-\varepsilon)s-1]Z_{t-1}-(q+s)Z_{t-1}^3-X_{t-1}+1}{Z_t}$$

Since term $\frac{1}{(1-\varepsilon)s}$ becomes very large when $(1-\varepsilon)s \to 0$, ratio $\frac{\Delta Z_t}{\Delta X_t} = \frac{Z_t - Z_{t-1}}{X_t - X_{t-1}} \to \infty$ except if $[q+(1-\varepsilon)s-1]Z_{t-1} - (q+s)Z_{t-1}^3 - X_{t-1} + 1 = 0$.

Let us consider the phase diagram (X_t, Z_t) or (Y_t, Z_t) with representation of unity on Y axis proportional to $(1-\varepsilon)s$.

- If $[q+(1-\varepsilon)s-1]Z_{t-1} - (q+s)Z_{t-1}^3 - X_{t-1} + 1 \neq 0$, the representation of the curve is given by a vertical line situated close to the extrema where $|Z_t| > 1.2$.

- On the other hand, if $[q+(1-\varepsilon)s-1]Z_{t-1} - (q+s)Z_{t-1}^3 - X_{t-1} + 1 = 0$, $X_{t-1} = [q+(1-\varepsilon)s-1]Z_{t-1} - (q+s)Z_{t-1}^3 + 1$ can be represented by a cubic in the phase diagram (X_t, Z_t).

Cubic function $X = [q+(1-\varepsilon)s-1]Z - (q+s)Z^3$ has a derivative : $[q+(1-\varepsilon)s-1] - 3(q+s)Z^2$ which cancels for two values of Z :

$$Z_1 = \sqrt{\frac{q+(1-\varepsilon)s-1}{3(q+s)}} \quad \text{et } Z_2 = -\sqrt{\frac{q+(1-\varepsilon)s-1}{3(q+s)}}$$

and since second derivative is $-6(q+s)Z$:

- The point (Z_1, X_1) where

$$X_1 = \left[q+(1-\varepsilon)s-1-(q+s)Z_1^2\right]Z_1 + 1$$

$$= \left[q+(1-\varepsilon)s-1-\frac{q+(1-\varepsilon)s-1}{3}\right]\sqrt{\frac{q+(1-\varepsilon)s-1}{3(q+s)}} + 1$$

$$= \frac{2}{3\sqrt{3}}\sqrt{\frac{q+(1-\varepsilon)s-1^3}{(q+s)}} + 1$$

is a local maximum.

- The point (Z_2, X_2) where

$$X_2 = \left[q+(1-\varepsilon)s-1-(q+s)Z_2^2\right]Z_2 + 1$$

$$= -\left[q+(1-\varepsilon)s-1-\frac{q+(1-\varepsilon)s-1}{3}\right]\sqrt{\frac{q+(1-\varepsilon)s-1}{3(q+s)}} + 1$$

$$= -\frac{2}{3\sqrt{3}}\sqrt{\frac{(q+(1-\varepsilon)s-1)^3}{(q+s)}} + 1$$

is a local minimum.

Let us notice, by coming back to (Y_1, Z_1) and (Y_2, Z_2), and since $X = (1-\varepsilon)sY$, that Y values becomes very large and would go out of the diagram if we had not taken for unity $\frac{1}{(1-\varepsilon)s}$ on Y axis.

5.3   ATTRACTION BASIN AND LIMIT

As a matter of fact, relaxation cycles of figures 4 to 7 are part of cubic appearing in figures 8 to 11.

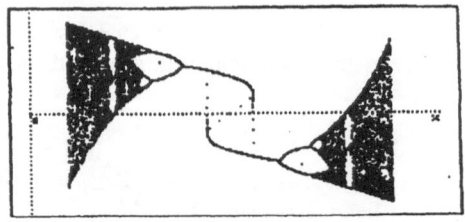

Figure 8: Cubic for $v = 1.5$

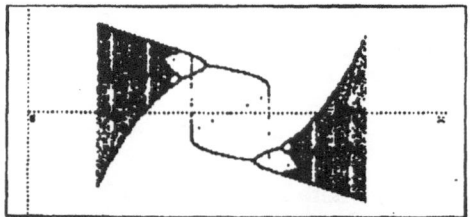

Figure 9: Cubic for $v = 1.75$

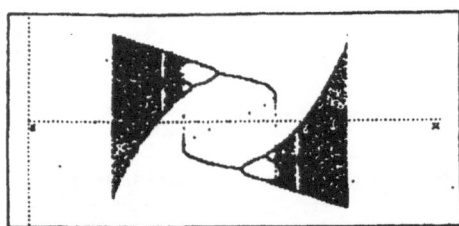

Figure 10: Cubic for $v = 1.87$

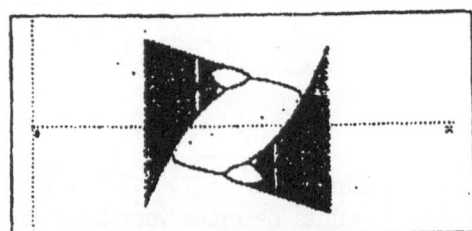

Figure 11: Cubic for $v = 1.12$

There are also the limits of the (Y_t, Z_t) wich are solution to (S) system, where initial value (Y_0, Z_0) is situated on the cubic or in its immediate neighbourhood. Cubic of figures 8 to 11 (and their immediate neibourhood) constitue the attraction basin for (S) model. Let us look, for instance, to figure 8 where cubic associated to $(1 - \varepsilon)s = 0.0001$ and $v = 1.5$. Starting from point $(Y_0, Z_0) = (1820, 0.5)$ situated near the left extremity of the cubic, we can see that the (Y_t, Z_t) of (S) model goes the right, first rather slowly (since there are some $Z_t < 0$ and some $Z_t > 0$, there having larger influence than the negative ones) then more quickly when all Z_t come to be positive and consequently Y_t are always increasing. This evolution continues to the neighbourhood of point (Y_1, Z_1) where $Y_1 = 11120$ is the cubic local maximum. In point (Y_1, Z_1), $\frac{dZ}{dY}$ tends to infinity and points (Y_t, Z_t) move to the half-straight line going down from (Y_1, Z_1). They are attracted by the inferior part of the cubic which they join at point C_1. In C_1, $Z_t < 0$ and points (Y_t, Z_t) decrease on the cubic to point (Y_2, Z_2) where $Y_2 = 8880$ and $\frac{dZ}{dY}$ tends again to infinity. From this point (Y_2, Z_2), points (Y_t, Z_t) of the model move to the half straight line up to point C_2 situated on the cubic and after that move on, keeping always on the cycle $(Y_1, Z_1)/C_1/(Y_2, Z_2)/C_2$ which stands as a limit cycle for (S) model.

If now v goes throught different values (say 1.75, then 1.87 and at last 2.12), keeping $(1 - \varepsilon)\,s = 0.0001$, one can see a decreasing amplitude of the cubic; on the other hand, the amplitude of the relaxation cycle -enriched successively by bifurcations and chaotic zones- goes increasing (cf. figures 9, 10, 11).

5.4   CUBIC'S FRONTIER

Let us come back to (S) system represented by a cubic with two symetrical parts around point $\left(\frac{1}{(1-\varepsilon)s}, 0\right)$. The upper part is limited to the rightside by point (Y_1, Z_1) local maximum of the cubic precedently defined ; on the left such there comes a chaotic zone with a frontier $\bar{Y}$ for v, s and δ given. It's obvious (cf. supra) that coefficient of Z_{t-1} in system (S), i.e. $q + (1 - \varepsilon)\,s = v - (1 - \varepsilon)\,s$ must be less then 3 ; if it is more than 3, Z_t goes to infinity. To calculate $\bar{Y}$, one draws the bifurcation diagram for the series $Z_t = [v - s + (1 - \varepsilon)\,s] - vZ_{t-1}^3 + b$ in a diagram with $b = -(1 - \varepsilon)sY + 1$ and as ordinates the stable equilibrium points of Z_t. b varies from 0 to 1 and the program stops when it comes to a value $\bar{b} < 1$ wich is reached for Z_t finite and Z_{t+1} getting out of the diagram. So that $\bar{Y} = \frac{1-\bar{b}}{(1-\varepsilon)s}$. The upper part of the cubic is defined for $\bar{Y} \leq Y \leq Y_1 = \frac{X_1}{(1-\varepsilon)s}$ and the (symetric) inferior part is defined for $\frac{X_2}{(1-\varepsilon)s} = Y_2 \leq Y \leq -\frac{2}{(1-\varepsilon)s} - \bar{Y}$.

So for $(1 - \varepsilon)s = 0.0001$ and $v = 1.75$, one gets the diagram of figure 12 and $\bar{Y} = 3400$ since $\bar{b} = 0.66$.

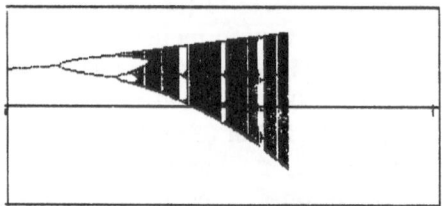

Figure 12: Bifurcation diagram for $v = 1.75$ $s = 0.001$ $1 - \varepsilon = 0.1$.

So the upper part of the cubics goes from $\bar{Y} = 3400$ to $Y_1 = 11890$. The lower part (symetrical from the upper part around point $\left(\frac{1}{(1-\varepsilon)s} = 10000\right)$ goes from $Y_2 = 8110$ to 20000 - $\bar{Y} = 16600$. The relaxation cycle, which appears as a limit cycle of model (S) goes from 8110 to 11890.

If the starting point is close to the relaxation cycle, the representation of the series (Y_t, Z_t) is given in the phase diagram by the relaxation cycle (in the paper, one choses $Y_0 = \frac{1}{(1-\varepsilon)s}$ and $Z_0 = 0.5$ when one wants that the successive points (Y_t, Z_t) to be situated on the relaxation cycle).

On the other hand, if the starting point of the cubic (Y_0, Z_0) is located to the left (resp. to the right) frontier of the relaxation cycle and after that to the cycle itself on which one travels indefinitely (figures 13 and 14).

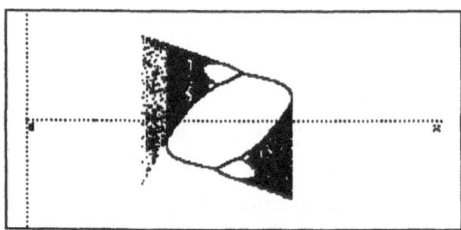

Figure 13: Cubic for $v = 2.1$, $(1 - \varepsilon)s = 0.001$ et $Y_0 = 570$

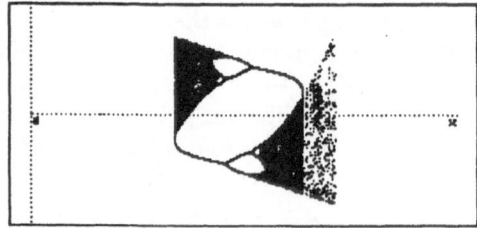

Figure 14: Cubic for $v = 2.1$ $(1 - \varepsilon)s = 0.001$ et $Y_0 = 1470$

5.5 RELAXATION CYCLE WITH THICK FRONTIERS

For $(1-\varepsilon)s$ given and negligible, it has been observed inside the relaxation cycle a chaotic zone growing in importance with v. If the starting point (Y_0, Z_0) is located on the superior part of the cubic, the series (Y_t, Z_t) is drawn (when arriving to the local maximum (Y_1, Z_1) where $\frac{dZ}{dY} \to \infty$), to the chaotic zone of the inferior part. Let us suppose that in this chaotic zone, one has some $Z_t > 0$ and some $Z_t < 0$, with relative influence quasi-equivalent, but slight avantage for $Z_t > 0$; consequently, terms (Y_t, Z_t) will stay during a "certain time" in a vertical strip $[Y_1 \bar{Y}_1]$ before going left to the local minimum (Y_2, Z_2) where $\frac{dZ}{dY}$ comes to infinity. Successive terms (Y_t, Z_t) will then linger, symetrically, in a vertical strip $[\bar{Y}_2 Y_2]$ before going right. So there will appear a relaxation cycle with two thick frontiers $[Y_1 \bar{Y}_1]$ and $[\bar{Y}_2 Y_2]$ (cf. figure 15).

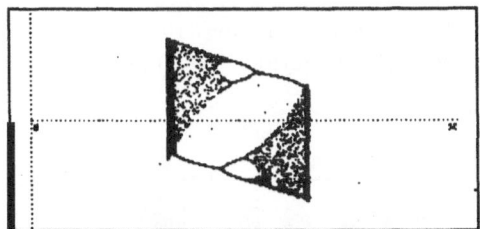

Figure 15: Relaxation cycle with thick frontier

We have precedingly given some privilege to diagram representation; now we shall call attention to time series of amounts and revenue change.

- The revenue level oscillates mainly between two values ("high" and "low" level) ; around these two levels, there are minor fluctuations and one could think that they are due to exogenous shocks , the length of "high" and "low" level stages are quite variable.

- Moreover, the stage of quick changes which appear on these diagrams have no connection with the "fast dynamics" which appeared in the cubic analysis. It is just the other way ; as is appears in the simulation, it is the stage of "slow" dynamics (in the phase diagram) which corresponds to the abrupt change in the time series. The two levels of revenue ("high" and "low") have variable length because the system wanders then through the "chaotic" part of the curve.

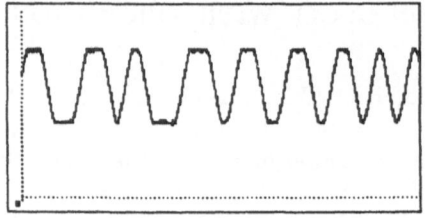

Figure 16: Variation of revenue level Y_t through time series

The above time-series is specially interesting ; first lenght of different stages are quite variable , second, there is abrupt change without any exogenous shock from one situation to the other:

- the chaotic behavior becomes clear when one examines the change in revenue level $Z_t = \Delta Y_t$; all precedent diagrams have been drawn in (Y_t, Z_t) coordinates; it is the revenue change of Z_t which suffers most important and unexpected changes. Revenue level Y_t is subjected to strong vibration with oscillations less important than these appearing for Z_t.

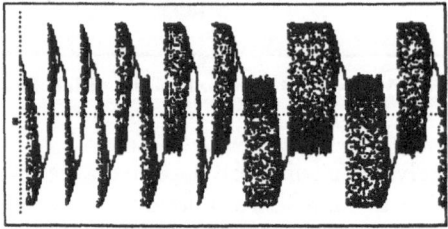

Figure 17: Variation of level-change Z_t through time

5.6   WHEN RELAXATION CYCLE DISLOCATES

Decomposition of the cubic: appearance of two limit strips
v continue to increase and the starting point (Y_0, Z_0) is situated in the upper hand of the cubic ; when the system arrives in the local maximum (Y_1, Z_1) where $\frac{dZ}{dY}$ is infinite, it is attracted into a chaotic zone of the inferior branch where some $Z_t > 0$ and some $Z_t < 0$ get mutually neutralized. If the starting point (Y_0, Z_0) is situated on the inferior branch of the cubic, close to its right frontier (which is a chaotic zone with some $Z_t > 0$ and some $Z_t < 0$), successive terms (Y_t, Z_t) go left until they enter into the vertical strips $[Y_1 \bar{Y}_1]$ where they will remain. This vertical strip $[Y_1 \bar{Y}_1]$ appears

as the limit of the series (Y_t, Z_t) solutions of model (S) when the starting point (Y_0, Z_0) is either on the upper part of th cubic with $Y_1 > Y_0 > Y_2$ either on the lower branch with $Y_0 > Y_1$

If the starting point (Y_0, Z_0) is situated either on the lower part of the cubic with $Y_2 < Y_0 < Y_1$ or on the upper part with $Y_0 < Y_2$, point (Y_t, Z_t) solutions of (S) have for limit a vertical strip $[\bar{Y}_2 \ Y_2]$ symetrical to the strip $[Y_1 \ \bar{Y}_1]$ around point $\left[\frac{1}{(1-\varepsilon)s}; 0\right]$.

So relaxation cycle has burst out ; limit of series (Y_t, Z_t) depends on initial conditions. For instance, for $(1 - \varepsilon)s = 0.0001$, we have such situation when $2.136 < v < 2.263$; if $v = 2.263$, $[Y_1, \bar{Y}_1] = [13623, 13716]$ and $[Y_2, \bar{Y}_2] = [6284, 6377]$.

Disappearance of relaxation cycle: divergent series (Y_t, Z_t).

If v still increased and with starting point $\left[Y_0 = \frac{1}{(1-\varepsilon)s}; 0.5\right]$, on the upper part of the cubic, series Y_t still increase, with $Z_t > 0$, to local maximum (Y_1, Z_1) where $\frac{dZ}{dY}$ tends to infinity.

Two possibility appears:

a) $Y_1 < \frac{2}{(1-\varepsilon)s} - \bar{Y}$ and system falls down to the chaotic zone of the inferior part where negative Z_t have less influence than positive Z_t.

Z_t will ultimately overshoot critical value $\bar{Z}$ and explodes, hence a very fast oscillating divergence of the series (Y_t, Z_t).

b) $Y_1 > \frac{2}{(1-\varepsilon)s} - \bar{Y}$; points (Y_t, Z_t) are no more drawn towards the cubic since they are too far from it. In such case, one has too very quick divergent oscillation (cf. figure 18).

As a matter of fact, we have just some points on straight lines coming from (Y_1, Z_1) or (Y_2, Z_2) before getting to a very quick divergent oscillation in series (Y_t, Z_t) (cf. figures 18 et 19).

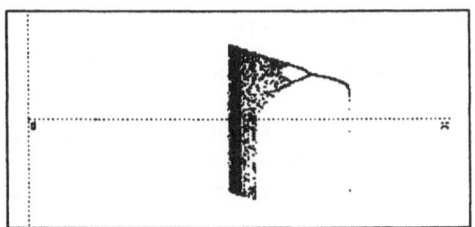

Figure 18: $v = 2.8 \ s = 0.01 \ (1 - \varepsilon)s = 0.1$ and $Y_0 = 1030$

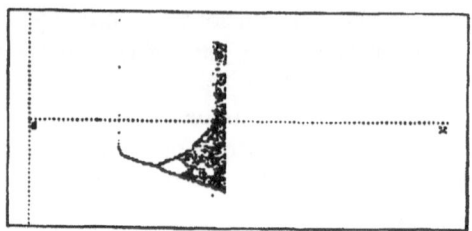

Figure 19: $v = 2.8$ $s = 0.01$ $(1 - \varepsilon)s = 0.1$ and $Y_0 = 960$

A central limiting strip for model (S)

With $(1 - \varepsilon)s = 0.0001$ and $v > 2.264$, relaxation cycle comes to dislocation. The starting point is $(Y_0, Z_0) = \left(\frac{1}{(1-\varepsilon)s}, 0\right)$ situated on a straight line going through the symetrical point of the cubic $\left(\frac{1}{(1-\varepsilon)s}, 0\right)$. In the close neighbourhood of the straight line $Y = \frac{1}{(1-\varepsilon)s} = 10000$, one has a dynamics quite similar to the one which appears in the bifurcation diagram of the series :

$$Z_t = [v - s + (1 - \varepsilon)s] Z_{t-1} - v Z_{t-1}^3$$

wich appeared in figure 2. So with $q = v - s + (1 - \varepsilon)s$, one knows that one obtains

- for $Q < 2$, a limit point

- for $2 < Q < 2.2357$, stable period 2 cycle

- for $2.2358 < Q < 2.287$, stable period 4 cycle

- for $2.2871 < Q < 2.3029$, stable period 8,16,...2n cycle

- for $2.303 < Q < 2.59809$, chaotic dynamic with $Z_t > 0$

- for $2.5981 < Q < 3$, chaotic dynamic with $Z_t > 0$ and $Z_t < 0$.

Obviously, if $Q < 2.59809$ and $(Y_0, Z_0) = (10000, Z_0)$ with $Z_0 > 0$, Z_t are all positive and Y_t increases to local maximum of cubic (Y_1, Z_1) ; according to the value of q, one has relaxation cycle, limit frontier strip or divergence.

When $Q > 2,59809$, one has negative $Z_t > 0$ and positive Z_t. If positive Z_t have more influence, the system will arive to the local maximum (Y_1, Z_1) and then diverges.

On the other hand, if positive and negative Z_t are mutually neutralized, the system stays in a narrow strip around the straight line $Y = \frac{1}{(1-\varepsilon)s}$. This is the situation represented by fig. 20 where $(1 - \varepsilon)s = 0.0001$, $v = 2.8$ and $(Y_0, Z_0) = (10020, 0.5)$.

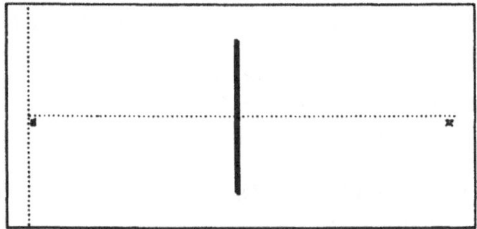

Figure 20: Central limiting strip $v = 2.8$, $(1 - \varepsilon)s = 0.0001$ and $Y_0 = 1020$

6 Analogy and difference between bifurcation diagram and cubic

It appears that the upper branch of the cubic looks like the bifurcation diagram of fig. 2 ; however, stable cycles and chaotic zone appear when one goes from right to left in the cubic but from left to right in the bifurcation diagram. There are more basic differences between the two diagrams. On a bifurcation diagram, one can observe a limit point, a stable cycle, or a chaotic evolution for a given value of $v - s$; all points of the cycle or of the chaotic dynamic are situated on the same vertical line drawn from point of abscisse $v - s$. On the other hand, stable cycles or chaotic dynamics are situated on parallel vertical lines distant by Z_t since $Y_t = Z_t + Y_{t-1}$. But since all these lines are squeezed one upon another, one has the impression that these points are on the same straight line (distance between two straightlines is about one when Y_t is about 10000).

A last difference must be noticed ; in the bifurcation diagram one has on each straight line the whole dynamics of (Z_t) for a particular value of $v - s$; in the upper (or lower) part of the cubic, one has the same value of $v - s$; in the upper (or lower) part of the cubic, one has the same value of $v - s + (1 - \varepsilon)s$; each vertical line corresponds to a particular Y_t and indicates just one point (Y_t, Z_t) of the series defined by (S) model.

References

[1] G. ABRAHAM-FROIS and E. BERREBI [1995], *Instabilité, cycles, chaos*, Economica

[2] C. CHIARELLA [1990], *Elements of a Non-Linear Theory of Economic Dynamics*, Springer-Verlag

[3] R. H. DAY and W. SHAFER [1986], "Keynesian Chaos", *Journal of Macro-economics*, 7.

[4] R. GOODWIN [1951], "The Non-Linear Accelerator and the Persistance of Business Cycles", *Econometrica*

[5] J. R. HICKS [1950], *A Contribution to the Theory of the Trade Cycle*, Oxford University Press.

[6] T. PUU [1991], *Non-Linear Economic Dynamics*, Springer-Verlag.

Credit and Financial Markets in Keynes Conception of Endogenous Business Cycles: an Interpretation

Richard ARENA and Alain RAYBAUT

1 Introduction

In his writings, John Maynard Keynes never developed a proper and systematic theory of the business cycle. However, the Treatise on Money and the General Theory include substantial indications on which it is possible to rely in order to explain the macroeconomic emergence and persistence of economic fluctuations. The only obstacle lies in the apparent contradictory elements which are present in Keynes's works: according to the post-Wicksellian viewpoint of the Treatise, the emphasis is put on the divergence between the natural and the market rates of interest and on its effects upon the demand for credit, while the General Theory stresses the role of human psychology (including on the stock exchange) in the determination of investment decisions.

This apparent contradiction is also present within modern Keynesian approaches. A good example is provided by the internal debate which, during the eighties, divided Post-Keynesian economists into two groups: the supporters of the theory of endogenous money supply (Moore 1988; Lavoie 1985) and the supporters of the liquidity preference theory (Kregel 1980, 87).

To begin with, our contribution will be devoted to the nature and reality of this contradiction. At a second stage, the use of a model of business cycles will show how it is possible to build a single framework allowing to cope with both components of the contradiction and, therefore, to explain why it was only apparent.

2   Banks, financial markets and business cycles

In the Treatise, as well as in the General Theory, Keynes relates the existence of business cycles to the adjustment mechanism between saving and investment. However, this relation takes in a different form depending on what book is considered.

In the Treatise, expansions and depressions correspond to situations where increases and decreases of the divergence between the natural and the market rates of interest. Thus, an initial impulse, causes for instance by a major innovation, increases the natural rate of interest. This increase induces, in turn, a wave of optimistic expectations and, therefore, an increase of the production level. This increase implies a rise of the demand for investment goods and thus of the demand for credit as well as greater employment and spending levels. Little by little, this expansion propagates and is generalised to all industries.

At first sight, this expansion might be seen as endless. Theoretically, banks are indeed able of creating an unlimited supply of credit:

" If we suppose a closed banking system, which has no relations with the outside world, in a country where all payments are made by cheque and no cash is used, and if we assume further that banks do not find it necessary in such circumstances to hold any cash reserves but settle inter-bank indebtness by the transfer of order assets, it is evident that there is no limit to the amount of bank money which the banks can safely create provided that they move forward in step " (Keynes, 1930, Vol I, p. 23).

Now, if this unlimited creation of credit occurs, that means that production and demand will increase. This remark shows the predominant role played by banks within the credit market. Accordingly, if the bank rate is defined by Keynes as " the effective rate for lending and borrowing which prevails in the market. " (Keynes 1930, Vol I, p. 179), this does not prevent however banks to exert their influence on its formation. This fact is essential since Keynes uses the term market rate of interest " for the complex of bank rate and bonds rate " (Keynes 1930, op. cit. p. 179) and " assumes that changes in bank rate affect the market rate of interest in the same direction." (Ibid. p. 180). This means that the banking system does not control the supply of money but the market rate of interest.(Keynes 1930, Vol Ii, p. 189):

" Apart from international complications which we are putting on one side for the present, there is no reason to doubt the ability of a central bank to make its short term rate of interest effective in the market. " (Keynes, 1930, Vol II, p. 325).

Therefore, the conditions of a Wicksell- type cumulative process are realised since the banking system is able to impose its own rate to the market and therefore, to regulate it. However, the idea of an unlimited credit supply is misleading if we consider the practice of the banking systems. It is indeed well-known that the mere existence of a central bank provides institutional constraints which commercial banks have to respect. Each of them is indeed supposed to hold a given volume of reserves in order to deal with that Keynes calls " the inevitable minor discrepancies which are bound to occur over short periods. " (Keynes, 1930, vol I, p. 24). Now, the amount of these reserves

" will depend partly on the habits of the depositors, as governed by the practices the country and of the period and by the class of business conducted by the particular bank's clients, and partly on the scale of the bank's business which for this purpose is generally measured by the amount of its deposits. " (Keynes, 1930, vol I, p. 25).

These constraints are institutional, behavioural, and also depend on social conventions, though they exert a macroeconomic effect on the volume of bank money which it is possible to create at a certain moment of time:

" We now perceive that there exists, not only the check on individual banks that they must keep step, but also a check on the banks as a whole. "

This is why " it is the aggregate of the reserve resources which determines the pace which is common to the banking system as a whole. " (Keynes, 1930, vol I, p. 25).

In other words, the amount of reserves determines the extent according to which the credit supply is limited. It prevents the permanence of a divergence between the natural and market rates of interest. At a given moment of an expansion phase, the ability of commercial banks to create credit can no longer be increased, since this increase will be contradictory with the necessity of maintaining a sufficient volume of reserves. As soon as this limit is reached, the supply of credit cannot extend further and the market rate of interest rises. The speed of reaction of the banking system therefore strongly depends on the nature and behaviour of the banking and financial institutions of the economy. It will contribute to determine the importance and the length of the business cycle.

The scheme provided by the General Theory differs substantially from the previous one.

The origin of the business cycle lies no longer in the existence of a divergence between the market and the natural rate of interest which the credit system allows to occur. It is now related to one of three " independent variables " of the General Theory, namely, the marginal efficiency of capital. This variable corresponds to the expected rate of profitability of new investments. Now, on one side, the expectations which govern the determination of this rate are fragile and volatile. On the other side, the life span of a capital good may not be anticipated with precision, especially in the industries in which there is technical progress. Last, the general conditions which determine the volume of investment belong to different types: " The considerations upon which expectations of prospective yields are based are partly existing facts which we can assume to be known more or less for certain, and partly future events which can only be forecasted with more or less confidence. Among the first may be mentioned the existing stock of various types of capital-assets and of capital-assets in general and the strength of the existing consumer's demand for goods which require for their efficient production a relatively larger assistance from capital. Among the latter are future changes in the type and quantity of the stock of capital-assets and in the tastes of the consumer, the strength of effective demand from time to time during the life of the investment under consideration, and the changes in the wage-unit in terms of money which may occur during its life. We may sum up the state of psychological expectation which covers the latter as being the state of long term expectations. " (Keynes, 1936, pp. 147-48).

Therefore, on the one hand, some determinants of long run expectations are fairly well known and their inclusion in the new investment decision-making process seems simple and clear-cut. On the other hand, for the remaining determinants, knowledge is replaced by ignorance, namely by a situation of " extreme precariousness of the basis of knowledge. " (Keynes, 1936, p. 149).

" Precariousness " is reinforced by the influence exerted by financial markets on the behaviour of investors. The division between owners and managers indeed allows financial agents to modify their investment decisions when they correspond to the acquisition of shares. Now, these acquisitions are not independent from the state of the stock exchange. The possibility for investors to reconsider their commitments increases the " precariousness " of investment expectations even more. We indeed know that " it is of the nature of organised investment markets, under the influence of purchasers largely ignorant of what they are buying and of speculators who are more concerned with forecasting the next shift of market sentiment than with a reasonable estimate of the future yield of capital-assets that, when disillusion falls upon an over-optimistic and over-bought market, it should fall with sudden and even catastrophic force. " (Keynes, 1936, p. 316).

Hence, in the General Theory, the emphasis is put on the variation of the marginal efficiency of capital: " The trade cycle is best regarded, I think, as being occasioned by a cyclical change in the marginal efficiency of capital, though complicated and often aggravated by associated changes in the other significant short-period variables of the economic system. " (Keynes, 1936, p. 313).

What about the long-run rate of interest? For Keynes, this rate is less unstable than the marginal efficiency of capital and, therefore, it only plays a complementary role in the emergence and development of the business cycle. The decrease of the efficiency of capital therefore implies the rise of the rate of interest and not the reversal. During the depression, as soon as the marginal efficiency of capital begins to decrease, agents try indeed to sell their inventories of goods and assets in order to avoid the decrease of their price. The level of liquidity preference therefore increases and, ceteris paribus, enhances the rise of the interest rate and hence reinforces the depression.

As a result, two different points of view seem to prevail. On the one hand, the existence of credit allows the possibility of a divergence between the natural and the market rate in favour of the latter. This divergence in turn underpins the phase of expansion. On the other hand, long-run expectations encourage investment decisions and, hence, expansion or depression phases.

Some authors have interpreted the opposition between these two points of view as the consequence of the respective different roles given by Keynesians or Post-Keynesians to banks and financial markets. This opposition must however be questioned.

Accordingly, we propose a formal framework allowing the analysis of the compatibility of these developments contained respectively in the Treatise and in the General Theory. We shall try to point out that, far from being contradictory, these developments can be combined in order to provide a more satisfactory explanation of the emergence and persistence of business cycles.

3 A synthetic model of endogenous fluctuations

In this section a synthetic business cycle model is developed. We first present the general framework of the model and derive its stationary state. Then we turn to the analysis of its local dynamics.

3.1 DETERMINATION OF THE LEVEL OF ACTIVITY AND OF THE EQUILIBRIUM ON THE MONEY MARKET

Let us start with a traditional Keynesian aggregated model of a closed economy where a single good is produced and where money and financial assets are held. Nominal income Y can be written as :

$$Y = C + I \tag{1}$$

where C and I respectively refer to the volume of national aggregated consumption and investment. In addition :

$$Y = wN + P \tag{2}$$

where P refers to the level of gross profits, and N refers to the level of employment. The nominal wage rate w is, in line with the Cambridge tradition, determined by a bargaining process that is exogenous to the model.

The consumption level, C, is given by:

$$C = cwN + c'\Pi^d \tag{3}$$

where c and c' refer respectively to the propensities to consume out of wages and out of distributed profits Pd. With:

$$\Pi^d = (1 - \beta)P, \text{ with } 0 < \beta < 1. \tag{4}$$

The employment level, N, is determined by the level of production Q and by labour productivity l.

$$N = \frac{Q}{l} \tag{5}$$

In addition, let us assume that the price of the produced good is given by means of an exogenous mark up m:

$$P = \frac{(1 + m)w}{l} \tag{6}$$

Thus, with a given a capital/output ratio v, the real sector of our model is completely described.

The description of the financial sector borrows from the analysis developed by Gallegati and Gardini (1991) and by Franke and Semmler (1991,92).[1]. One main difference can be found in the fact that we consider two distinct interest rates. The first rate $\hat{\imath}$, is parametrically fixed by the banks and refers to the " bank rate " of the Treatise. The second one, i, is detemined by the equilibrium on the assets market and refers to the long-term rate of the General Theory.

Let us assume for simplicity that the money supply M^s is constant and exogenous. The demand for money has three components: a transactional and precautionary element TR, a speculative component consistent with the notion of liquidity preference of the General Theory, SP, and finally a component related to the demand for finance Exfi.[2].

Accordingly, assume that:

$$TR = kY, \tag{7}$$

where $0 < k < 1$ is the reciprocal of the velocity of money

$$SP = SP(i) \tag{8}$$

$$Exfi = \dot{D} \tag{9}$$

---

[1]Hence, for a detailed presentation of the financial sector see Gallegatti and Gardini (1991) and Franke and Semler (1992).

[2]As we know, the " finance motive" historically appears after the publication of the General Theory.

It is now time to indicate how the real and financial sectors are connected. from a Keynesian perspective, this link is to be found in the financing of investment activity. From this standpoint, three sources of financing are taken into account in the model: retained net profits A, debt (bank loans) $\dot{D}$, and finally access to financial markets $\dot{M}$ .Thus we obtain :

$$I = A + \dot{D} + \dot{M} \qquad (10)$$

Let us assume that the internal financing constraint is a proxy for the ability of firms to fulfill their debt commitments. Hence, according to Kalecki's increasing risk principle, new external finance raised by the firms is proportional to net profits, i.e. net cash flow A. We have :

$$\dot{D} = (b - 1)A \qquad (11)$$

with $b > 1$, and where A is given by

$$A = \beta\Pi - iD \qquad (12)$$

The investment function we have retained is the following :

$$I = bA + a(r^e - i) \qquad (13)$$

where $a > 0$ and $b > 1$ are two real positive parameters. Thus, investment decisions depend both on net cash flow A and on the difference between the expected profit rate and the current market interest rate. Consequently (b-1) refers to the weight of financial constraints undergone by firms (Gallegati et Gardini 1991)[3].

Assume, moreover, that the expected profit rate is given by the ensuing relation:

$$r^e = R(d, X) \qquad (14)$$

where d refers to the debt/capital ratio and X refers to the state of confidence. Let us suppose that R is a continuous derivable positive function satisfying $R'_x > 0$.

According to Minsky (1980, 82), the evolution of the leverage ratio D/K during the cycle precedes an increase (respectively a decline) of the financial fragility of the economy. A small D/K ratio is first associated with high expected profits, i.e. $R'_d > 0$. Progressively, a rise in D/K signals a future deterioration of profitability, hence R'_d becomes negative.[4].

Since the real positive variable X refers to the notion of " state of confidence " of the General Theory mentioned in the introduction of the paper,

---

[3]A simple combination of equations (10), (11) and (13) shows that the amount of investment financed by the financial market is given by $\dot{M} = a(r^e - i)$.

[4]Thus, the function $R(d, X)$ can be specified as follows: $R(d, X) = \alpha d + \beta d + \delta X$, with $\alpha < 0, \beta > 0$ and $\delta > 0$.

investment decisions depend positively on X. As we may recall, fluctuations of the state of confidence also act upon aggregate activity through the channel of financial markets. However, in this model, fluctuations of investment decisions do not rely on an hypothesis of volatility of confidence on financial markets, volatility transmitted to the real sector. Accordingly, as noticed by Franke and Semmler (1992), " should that have been the case, the economy might be seen as being too much dependent on the psychology of speculators " (p.331). In line with the General Theory, it is possible to consider that real and financial sectors are simultaneously affected by changes of the state of confidence, " the impact on investment demand being the most direct " (ibid.). From this perspective, in our model the interplay of the state of confidence and of the financial sector is understood in an indirect way through the determination of expected profits and of the interest rate.

Let us finally consider the condition for equilibrium on the financial and money markets. The money market is characterized by two elements. On the one hand, the financial gap, i.e. the demand for external finance, is a fraction of the demand for money. On the other hand, the amount of liquidity created by the banking system is, for a given banking interest rate and without credit rationing, driven by the firms' demand for finance. Recent endogenous business cycles or cyclical growth models (Gallegati and Gardini, 1991, Foley, 1986, 87, Assada and Semmler, 1992, Semmler and Sieveking, 1993), adopt the same perspective. We obtain :

$$M^s = M^d = M \Leftrightarrow M - kY - \dot{D} - SP(i) = 0 \qquad (15)$$

Replacing Y and $\dot{D}$ by their respective values in K and D, we have:

$$M - C_0 K + \hat{i}(b-1) - SP(i) = 0, \text{ with } C_0 = \frac{w(k(1+m) + \beta m(b-1))}{lv} \qquad (16)$$

Relation (15) can be rewritten as :

$$i = L(K, D)$$

Hence, this type of LM curve with $L'_K < 0$ and $L'_D > 0$ (cf. Annex 1) gives the equilibrium or market interest rate as a function of the capital stock and of the stock of debt.

The real and financial sectors of our model having been described, it is now possible to determine the stationary state of the economy and to turn to the analysis of its dynamical properties.

3.2 ENDOGENOUS BUSINESS CYCLES

From the equations given in the previous sections, we derive a nonlinear dynamical system with three state variables : the capital stock K, the stock of debt D, and the state of confidence X. The dynamics of capital stock and of debt is given by:

$$\dot{K} = I \tag{17}$$

$$\dot{D} = (b - 1)A \tag{18}$$

As we know, variations in the psychological variable representing investors' confidence are an essential ingredient in Keynes' conception of macroeconomic instability in the General Theory. We give a somewhat simplified, but sufficient for our purpose representation of the evolution of that variable. Let us simply assume that its movement is given by the following relation:

$$\dot{X} = H(X) \tag{19}$$

where H is a continuous function defined for all $X > 0$, satisfying $H(X^*) = 0$ and $H_X'^* \neq 0$ for $X^* > 0$[5].

In line with Keynesian business cycles models based on fluctuations of the state of confidence which usually refer to the General Theory's Notes on the Trade Cycle and to the 1937 QJE' article (Woodford 1989, Caminati 1989), the evolution of the state of confidence, which is basically a psychological variable is autonomous. Function H captures the idea that this variable will fall (rise) if a deterioration (improvement) of present conditions is expected. Let us for instance consider the upswing period of a cycle. Investors who are optimistically minded should expect confidence to increase more or less steadily. Conversely, after the crisis, during the downswing period, investors should expect confidence to decrease. However, as noticed by Boyd and Blatt (1988), as time goes on, " the memory of the last panic becomes less vivid, and investors become more willing to look farther into the future with confidence that their predictions and expectations will be met " (p. 65), bringing about a reversal of opinion. [6].

From equations (17)-(18)-(19) a unique stationary state can easily be derived. We obtain the following result :

Proposition1: existence of a stationary state

---

[5]We can for instance specify this function as follows: $H(X) = \theta_1 X2 + \theta_2 X$, with $\theta_1 < 0$ and $\theta_2 > 0$.

[6]A complementary approach consists, in line with Kalecki (1937) and Minsky, in taking into account the role of increasing risk and of financial fragility. As noticed by Franke and Semmler (1991), this risk embodied in a function of the debt capital ratio D/K exerts negative effects on the state of confidence. Thus, in that case we have $H_d' < 0$.

Assume that $M > SP(i^*)$ where i^* refers to the market interest rate evaluated at the stationary state. System (18)(19) then has a unique strictly positive stationary point $E = (K^*, D^*, X^*)$. We obtain :
$K^* = \frac{-lv(SP(i*)-M)}{kw(1+m)}$, $D^* = f_K^*$ where $f = \frac{\beta mw}{lvi}, X = X^{*7}$

Proof:

- On the one hand, it is clear that $\dot{D} = 0$ implies $A = 0$. Hence, $D^* = \phi K$, with

$\phi = \frac{\beta mw}{lvi}$. At the stationary point, the debt/capital ratio, D/K, is constant and equal to ϕ.

- On the other hand, $\dot{K} = 0$ implies $\frac{-b}{a} A = r^e - i$. Thus, at the stationary point we have $r^e - i = 0$, the expected profit rate is equal to the market rate. Consequently, $r^e = R(f, X^*) = i^*$. Now, the equilibrium condition in the money market implies:

$$M - SP(i^*) = C_1 K, \text{with } C_1 = \frac{wk(1+m)}{lv}. \tag{20}$$

Hence, it can easily be deduced that $K^* = \frac{-lv(SP(i^*)-M)}{kw(1+m)}$, which means that $K^* > 0$ if and only if $M > SP(i^*)$.]

The remainder of the paper is devoted to the analysis of local properties of the dynamical system (17), (18), (19). Consequently, we consider a linearization of the model in a neighbourhood of the stationary state $E = (K^*, D^*, X^*)$. Thus, the characteristic equation can be written as

$$(J_{33} - z) \left(z^2 - (J_{11} + J_{22}) z + J_{11} J_{22} - J_{21} J_{12}\right) \tag{21}$$

where the J_{ij}'s refer to the elements of the Jacobian matrix given in Annex 2.

It is obvious that (20) has a positive or negative real root $z^3 = H'X^*$. The two other roots are then real or complex. We shall first give a necessary and sufficient condition for the existence of two complex roots.

Proposition 2: existence of complex roots
The characteristic equation (20) evaluated at the stationary state, has two complex roots if $R'_d < R'_d < R_{\overline{d}}$, where $R'\underline{d} < 0$ et $R_{\overline{d}} > 0$. In other words, the model has two complex roots if, in a neigbourhood of the stationary state, the reaction of the expected profit rate to the debt capital ratio D/K is not too high.

---

[7]With the specification of the function H(X) mentioned above, we get $X^* = -\frac{\theta_2}{\theta_1} > 0$

[Proof:

The model admits complex roots if: $(J_{11}^* - J_{22}^*)^2 + 4J_{21}^* J_{12}^* < 0$

Replacing the element of the Jacobian matrix by their respective values evaluated at the stationary state given in Annex 2, we obtain : $(J_{11}^* - J_{22}^*)^2 = (C_2 R_{d*}' + C_3)^2 \Leftrightarrow 4J_{21}^* J_{12}^* = 4(C4 R_{d*}' + C_5)$

where coefficients C_2, C_3, C_4, C_5 are the following:

$C_2 = \frac{-ak(1+m)}{lv(M - SP(i*))} < 0$

$C_3 = i(b-1)\left(1 - \frac{a\phi}{SP'(i^\bullet)}\right) + b_i \phi - \frac{awk(1+m)}{lvSP'(i^\bullet)}$

$C_4 = \frac{(b-1)ak(1+m)\phi}{lv(M - SP(i*))} > 0$

$C_5 = \hat{i}(b-1)((b-1)a - b)\phi$

Then, complex roots exist if:

$C_{22} R_d'^{*2} + (2C_2 C_3 + 4C_4) R_d'^* + (C_{32} + 4C_5) < 0$

Assume now that $C_{32} + 4C_5 < 0$, we obtain :

$\underline{R_d'}, \overline{R_d'} = \frac{-(C_2 C_3 + 2C_4) \pm \sqrt{(C_2 C_3 + 2C_4) - C_2(C_3 + 4C_5)}}{C_2}$

Consequently, the model has two complex roots if $\underline{R_d'}, < R_d' < \overline{R_d'}, .$]

We then make two assumptions related to the banking rate, $\hat{i}$, and on the state of confidence, X, in a neighbourhood of the stationary state which ensure the existence of a Hopf bifurcation from the left.

Assumption 1

We suppose that $r > \hat{i}$. The bank rate $\hat{i}$ charged by banks is lower than the actual profit rate r:

$$\hat{i} < \frac{\beta m w}{lv}\left(1 - \frac{a}{SP(i^*)}\right) \Rightarrow \hat{i} < br\left(1 - \frac{a}{SP(i^*)}\right) \tag{22}$$

Assumption 2

In a neighbourhood of the stationary state, we have $H_X'^* \neq 0$. We assume moreover that $H_X'^* \neq 0$. That around the stationary state, the state of confidence is sensitive to small perturbations in opinion, but a feedback effect prevails. Thus, assumption 2 means that fluctuations in the psychological variable are not the main destabilizing factor in the model. [8]

Under assumptions 1 and 2 we obtain the following result.

Proposition 3: existence of a Hopf bifurcation

Choose the parameter $\gamma = (b-1)$ which measures the weight of the internal finance constraint, as a bifurcation parameter. Then a Hopf bifurcation takes place at $\gamma = \gamma^*$.

[Proof:

---

[8]It is obvious that this condition is true when the function $H(X)$ is specified as in note 3 above.

This result relies on the Andronov-Hopf bifurcation theorem.

- Proposition 1 tells us that (20) has a pair of complex roots $z_1, z_2 = a(\gamma) \pm ib(\gamma)$ and a real negative root z_3.(Assumption 2)

- For the critical value g*, (20) has a pair of complex roots crossing the imaginary axis at non zero speed with: . $(1 - \gamma*)\frac{\partial \, \mathrm{Re}(z)}{\partial(1-\gamma)} \neq 0$.

The real part of z_1 and z_2 can be written :

$a(\gamma) = \gamma B_1 + B_2$

where coefficients B1 and B2 are the following:

$B_1 = -\hat{\imath} + \frac{\beta m w}{lv}(1 - \frac{a}{SP(i*)})$

$B_2 = \frac{\beta m w}{lv} + \frac{a\beta m w k(1+m)R'_{d*}}{i(lv)^2(SP(i*)-M)} - \frac{awk(1+m)}{lvSP'(i*)}$

It is easy to verify that $B1 > 0$ if $\hat{\imath} < br(1 - \frac{a}{SP(i*)})$ (Assumption 1).

Thus, a Hopf bifurcation from the left takes place for $g = g*$. (N.V. Tu 1994 p. 202)]

When this bifurcation is subcritical, the stationary state (K^*, D^*, X^*) is an unstable focus for $g > g*$, stable for the critical value and an unstable cycle exists for $g < g*$ [9]. In that case, the model exhibits, using Leijonhuvfud's words, a " corridor of stabilty " property" [10] Conversely, when the bifurcation is supercritical the stationary state is a stable focus for $g < g*$ which becomes unstable for $g = g*$. A stable cycle appears for $g > g*$ and the economy experiments endogenous persistant cycles.

Hence, it appears that the financial structure of investment, especially the weight of the internal finance constraint g, plays a crucial role in the dynamics and prevents the development of a cumulative process.

4 Concluding remarks

In this contribution, the acknowledged contention of the existence of two different, even contradictory, approaches of economic fluctuations in Keynes' writings has been challenged. Indeed, the nonlinear model of endogenous fluctuations we have proposed has shown that, for Keynes, business cycles can be simultaneously explained by the credit market and by expectations (including on the stock market) of investment.

On the one hand, the state of long-term expectations plays an important

---

[9]The bifurcation-type, and therefore the stability criterium of the cycle, depends on the sign of a coefficient S. If $S < 0$, the bifurcation is supercritical, while the bifurcation is subcritical when $S > 0$. However, computation of S composed of seconds and third derivatives orders of the Taylor approximation of the system could be messy, even in the simpler case.(Guckenheimer and Holmes 1983 p.154-56).

[10]The paternity of this remark concerning this type of Hopf bifurcation dynamics is due to Franke and Semmler (1992) op. cit. p.344.

role in the analysis of persistent economic instability. If the state of confidence actually exerts a direct positive effect on investment decisions, its mere autonomous evolution is not however the unique cause of instability. Indeed, the emergence of the cyclical behaviour of the economy is greatly explained by the financial structure of investment. So, the explanation of business cycles outlined in the General Theory is enriched. On the other hand, the interplay of the two interest rates of the Treatise is also taken into account. The first rate, the assets yield, is determined by equilibrium conditions on the financial market, while the other is the loan rate fixed by the banks. The latter is a parameter imposed by the banking system which constrains the cyclical dynamics.

Our findings connect in many respects to New Keynesian Macroeconomics. In this literature, credit and financial markets do explain a substantial part of economic fluctuations. Nevertheless, our approach based on an endogenous business cycles model relates more closely to Keynes's view on the inherent tendency to instability of market economies.

5 Appendix

Annex 1

Partial derivatives of the equilibrium interests rate are the following:
$i'K = \frac{C1}{SP(i)} < 0$
$i'D = \frac{-\hat{i}(b-1)}{SP(i)} > 0$
Partial derivatives of the expected profit rate are :
$r^{e\prime}K = -\frac{D}{2K}dR$
$r^{e\prime}D = \frac{1}{K}dR$
$r^{e\prime}X = \hat{R}'X$

Annex 2

The elements of the jacobian matrix J evaluated at the stationary state are:
$J_{11}^* = a(r_{K*}^{e\prime} - i'_{K*}) + b\frac{\beta mw}{lv}$
$J_{12}^* = a(r_{D*}^{e\prime} - i'_{D*}) - b\hat{i}$
$J_{13}^* = ar_{X*}^{e\prime}$
$J_{21}^* = (b-1)\frac{\beta mw}{lv} > 0$
$J_{22}^* = -\hat{i}(b-1) < 0$
$J_{23}^* = 0$
$J_{31}^* = 0$
$J_{32}^* = 0$
$J_{33}^* = H'_{x*}.$

References

[1] ARENA R. and RAYBAUT A. (1995), "Cycles et croissance, un point de vue né o-kaldorien", *Revue Economique*, Nov-Dec.

[2] ASADA T. and SEMMLER W. (1992), "Growth, Finance and Cycles", Working paper New School for Social Research, New York.

[3] BOYD I. and BLATT J.M. (1988), *Investment, Confidence and Business Cycles*, Springer Verlag.

[4] CAMINATI M. (1989), "Cyclical Growth and Long-Term Prospects", *Political Economy*, Vol. 5 N° 2 p.107.

[5] DELLI GATTI D., GARDINI L. and GALLEGATI M. (1993), "Investment Confidence, Corporate Debt and Income Fluctuations", *Journal of Economic Behavior and Organization*, 22, October, p.154-161.

[6] FRANKE R. (1992), "Stable, Unstable, and Persistant Cyclical Behaviour in a Keynes-Wicksell Monetary Growth Model", *Oxford Economic Papers*, 44, p.242-56.

[7] FRANKE R. and SEMMLER W. (1991), "Expectations, Dynamics, Financing and Business Cycles.", in *Profits, Deficits and Instability* edited by D.B Papadimitriou MacMillan

[8] GALEGATI M. and GARDINI L. (1991), "A Non Linear Model of Business Cycles with Money and Finance", *Metroeconomica*, Vol 42 N° 1 p.1

[9] GUCKENHEIMER J. and HOLMES P. (1986), *Non Linear Oscillations Dynamical Systems and Bifurcations of Vector Fields*, Springer Verlag

[10] JARSULIC M. (1993), "Complex Dynamics in a Keynesian Growth Model", *Metroeconomica*, Vol. 44 Fev. p. 43.

[11] KALECKI M. (1937a), "A theory of the Business Cycle", *Review of Economic Studies*, Feb. p. 77.

[12] KALECKI M. (1937b), " The Principle of Increasing Risk", *Econometrica*, Nov p. 441.

[13] KEYNES J.M. (1930), *A Treatise on Money*, Vol 1 et 2, in The Collected Writings, vol 5 and 6, MacMillan 1971.

[14] KEYNES J.M. (1936), *The General Theory of Employment, Interest and Money*, in The Collected Writings, vol 7 MacMillan, 1973.

[15] KEYNES J.M. (1937), '' The General Theory of Employment '', *Quarterly Journal of Economics*, Vol 51 p.209

[16] KREGEL I. (1980), " Expectations and Rationality within a Capitalist Framework ",. Paper presented at the American Economic Association Meeting, Denver.

[17] KREGEL I. (1987), " Irving Fisher, Great-Grand Parent of the General Theory: Money, Rate of Return, Overcosts and Efficiency of Capital ", mimeo, Jonhn Hopkins University, Bologna Center.

[18] LAVOIE M. (1985), " Credit and Money: the Dynamic Circuit, Overdraft economies and Post Keynesian Economics ", in M. JARSULIC ed. *Money and Macropolicy*, Kluwer publisher, Boston.

[19] MOORE B. (1988), *Horizontalits and Verticalits: the Macroeconomics of Money and Credit.*CUP, Cambridge.

[20] MOORE B.(1988), " The Endogenous Money Supply ", Journal of Post-Keynesian Economics, vol 10, N° 3.

[21] PUU N.V. (1994), *Dynamical Systems*, Springer Verlag

[22] SEMMLER W. (1986), "On Nonlinear Theories of Economic Cycles and Persistence of Business Cycles", *Mathematical Social Science*, 12(1) p.47-76

[23] SEMMLER W. (1989), ed. *Financial Dynamics and Business Cycles: New Perspectives*, M.E. Sharpe, New York.

[24] SEMMLER W. and SIEVEKING M. (1993), "Nonlinear Liquidity Growth Dynamics", *Journal of Economic Behavior and Organization*, 22, October. p. 189-208.

[25] WOODFORD M. (1989), "Finance, Instability and Cycles ", in W. Semmler ed *Financial Dynamics and Business Cycles* M.E Sharpe

Part 4

Methodological Issues

Business Cycles, Chaos and Predictability

Alfredo MEDIO[1]

1 Business Cycles and Chaos: An Introduction

A scanty observation of the time series of most variables of economic interest, such as the price of an individual commodity or the exchange rate between two currencies, shows the presence of bounded and more or less regular fluctuations, with or without an underlying trend. Even more interestingly, this oscillating behaviour seems also to characterize the aggregate activity of industrialized economies, as represented by their main economic indicators like GNP, global production, consumption and investment, employment or exports. Economists have long been concerned with the explanation of this phenomenon. The literature on the subject is enormous and the number of different theories equally vast. However, if we restrict ourselves to the *mathematical* investigation of economic fluctuations, we observe that two basic, mutually competing approaches have dominated this area of research in modern times.

The origin of the first approach - which we shall label "econometric approach to business cycles" - may be traced back to the seminal works of Slutsky (1927), Yule (1927) and Frisch (1933) and was later developed and given the status of orthodoxy by the works of the Cowles Commission in the 1940s and 1950s.

The fundamental idea of the econometric approach is the distinction between *impulse* and *propagation mechanisms*. In the typical version of this approach, serially uncorrelated shocks (the impulse mechanism) affect the relevant variables through distributed lags (the propagation mechanism), leading to serially correlated fluctuations in the variables themselves[2].

---

[1]The author gratefully acknowledges the financial support of the Italian Ministry of the University (M.U.R.S.T.) and the National Council of Research (C.N.R.).

[2]For completeness's sake, among the impulse-propagation models of the cycle, one should distinguish between those in which random external events affect economic "fundamentals" (essentially, tastes and technology), and those in which

As Slutsky showed, even simple linear non-oscillatory propagation mechanisms, when excited by random, structureless shocks, can produce output sequences which are qualitatively similar to certain observed macroeconomic cycles.

The ability of the econometric approach to provide an explanation of business cycles was called in question largely on the ground that explaining fluctuations by means of random shocks amounts to a confession of ignorance. An alternative approach - which we shall label "nonlinear disequilibrium (NLD)" - was then developed by a school of economists who, somewhat misleadingly, was associated with the name of Keynes. The basic idea of these authors was that instability and fluctuations are essentially due to market failures and consequently they must be primarily explained by deterministic models, i.e., by models where random variables play no essential role. Classical examples of such models can be found in the works of Kaldor (1940), Hicks (1950), Goodwin (1951). Mathematically, these models were characterized by the presence of nonlinearities in certain basic functional relationships of the system and lags in its reaction-mechanisms. The typical result was that, under certain configurations of parameters, the equilibrium of the system can loose its stability, giving rise to a stable periodic solution (a "limit cycle"), which was taken as an idealized description of self-sustained real fluctuations, with each boom containing the seeds of the following slump and vice versa. The NLD approach to the analysis of business cycles was very popular in the forties and fifties, but its appeal to economists seems to have declined rapidly thereafter and a recent, not hostile textbook of macroeconomics (Blanchard and Fischer, 1987, p. 277) declares it 'largely disappeared'.

The reasons for the crisis of the Keynesian style of theorizing and the related NLD theories of the cycle are manifold, not all of them perhaps pertaining to scientific reasoning, and a full investigation of this interesting issue is out of the question here. However, there exist two fundamental criticisms, raised against the NLD approach mainly by supporters of the rational expectations hypothesis, which are relevant to our discussion and can be briefly summarized thus:

• In the NLD models, agents' expectations, either explicitly modelled, or implicitly derived from the overall structure of the model, are, under most circumstances, incompatible with agents' "rational" behaviour.

• The NLD approach has been "refuted" by empirical observation, as time series generated by the relevant models do not agree with available data. For example, so the argument runs, a time series generated by a model characterized by a stable limit cycle will have a power spectrum exhibiting a sharp peak corresponding to the dominant frequency, plus perhaps a few

---

those events directly change only agents' expectations. The latter case has been extensively studied in recent years in the economic literature under the label "sunspots".

minor peaks corresponding to subharmonics. On the contrary, aggregate time series of actual variables would typically have a broad band power spectrum, often with a predominance of low frequencies (see, for example, Granger and Newbold, 1977).

The first of these criticisms can be best appreciated by making reference to the original formulation of the rational expectations hypothesis. In Muth's own words, "the expectations of firms (or, more generally, the subjective probability distribution of outcomes) tend to be distributed, for the same information set, about the prediction of the theory (or the "objective" probability distributions of outcomes)" (see, Muth, 1961, p. 316). If expectations were not rational in the sense defined above – so Muth's argument continues – 'there would be opportunities for economists in commodity speculation, running a firm, or selling the information to the present owners' (see, Muth, *loc. cit.*, p. 330). This argument does have some validity if the outcomes of the dynamical system under consideration can be accurately predicted once the 'true' model is known, e.g., if the outcome is periodic. In this case, if agents are 'rational', fluctuations can be explained only by exogenous random shocks[3].

However, if the theory implies chaotic, unpredictable dynamics of the system, the rational expectations argument loses much of its strength and non-optimizing rules of behaviour – such as adaptive reaction mechanisms of the kind assumed by the NLD models, or 'bounded rationality' *à la* Simon – might not be as irrational as they may seem at first sight. At any rate, the presence of chaos makes the hypothesis of costless information and the infinitely powerful learning (and calculating) ability of economic agents, implicit in the perfect foresight–rational expectations hypothesis, much harder to accept.

Equally strong reservations can be raised in relation to the second criticism mentioned above. It is well known that deterministic chaotic systems can generate output qualitatively similar to the actual economic time series, e.g., they can have a similar power spectrum.

None of these broad considerations can be used as a conclusive argument that business fluctuations are actually the output of chaotic deterministic systems. However, they strongly suggest that, in order to describe complex dynamics mathematically, one does not have necessarily to make recourse to exogenous, unexplained shocks. The alternative option - the deterministic description of irregular fluctuations - provides economists with new research opportunities undoubtedly worth exploiting.

---

[3]Even when the outcome is periodic, however, if the periodicity is long and the time path very complicated, one may question the idea that well-informed real economic agents would actually forecast it correctly: all the more so if the outcome is quasi periodic, possibly with a large number of incommensurable frequencies.

2 Stochastic Behaviour of Deterministic Systems

Although a universally accepted and comprehensive characterization of chaos is still lacking, in this paper we shall adopt the definition suggested in a recent conference on chaos (Royal Society, London, 1986), namely, 'stochastic behaviour occurring in a deterministic system'. Broadly speaking, a system is said to be deterministic when it comprises no exogenous random variables. On the other hand, the *observable* behaviour of a dynamical system is called stochastic when the transition of the system from one state to another can only be given a probabilistic description as happens for truly random processes, e.g., the outcome of spinning a roulette wheel. The meaning of these rather intuitive considerations will be made clearer, we hope, in the following pages. Indeed, much of this paper is devoted to giving a precise characterization to the basic notions of stochastic behaviour of deterministic systems, chaos, unpredicatbility and other related concepts.

Before carrying further our investigation, however, we would like to pause and briefly consider some interesting preliminary questions. First of all, is deterministic chaos – or if you prefer, stochastic behaviour of deterministic systems – an intriguing but rare event, or is it a general and ubiquitous occurrence? A broad answer to this question is: deterministic chaos is not a rare event in the sense that, as is well known, it can be found generically in "small" and "simple" dynamical systems e.g., non-invertible one-dimensional maps, or three-dimensional systems of differential equations, characterized by a simple quadratic nonlinearity.

A more specific, and economically more interesting question is the following: how relevant is chaos to economic theory and how common is chaotic behaviour in real economies? To the first part of this question, we can give a straight and definite answer, as follows. If we consider the two basic dynamical models of the prevailing economic theory - i.e., optimal growth and overlapping generations model - we know that, generically, their output can be chaotic (see, for example, Boldrin and Montrucchio, 1986; Deneckere and Pelikan, 1986; Medio, 1992; Medio and Negroni, 1996). "Generically" here means that chaos occurs for parameter constellations that we cannot exclude *a priori* on the basis of the economic "first principles". Hence, if one takes those models seriously as most members of the economic profession do, chaos must be considered a very real possibility. A similar conclusion is reached if we consider non-equilibrium dynamical models inspired by the Keynesian tradition.

It is more difficult to give a neat answer to the second part of the question above. Here we have two possible strategies. On the one hand, we could make recourse to the analysis of economic data, trying to find empirical evidence of chaos. There is a vast literature on this subject but we cannot discuss it here in any detail. In our opinion, no general consensus has yet been reached, except on the fact that it is unlikely that economic data may be represented by low-dimensional chaos (say, an orbit on a three-dimensional chaotic attractor). A different, not necessarily alternative route is to try and "calibrate" a theoretical model, i.e., estimate the parameter values deemed to be more realistic and verify if they correspond to chaotic behaviour. Although this is a reasonable procedure in principle, we are rather skeptical about its validity in practice. A serious "calibration" would require theoretical models conceived in such a way that they can be empirically tested. Unfortunately, economic models - or at least those describing equilibrium dynamics - have a normative rather then positive character, as they are constructed to described *what rational agents should be doing rather than what real economic agents actually do.*

At any rate, in this paper we shall not develop this point any further but shall try instead to clarify some general theoretical issues whose proper understanding is, in our opinion, crucial for a correct assessment of the relevance of chaos for economics. In particular, we shall investigate the basic question of predictability. Rather than discussing the issue in full generality, we shall consider its simplest possible formulation containing all the interesting features that concern us here. In this way, we shall obtain a greater transparance with a smaller amount of technicalities.

3   The Simplest Representation of Chaotic Dynamics

From the mathematical literature on chaos, we know that the simplest deterministic model capable of stochastic behaviour can be represented by a map (a difference equation) such as

$$x_{t+1} = f(x_t) \tag{1}$$

where x denoted a real scalar variable and f is a "one-hump" function, such as a quadratic polynomial. From the economic literature (see, for example, the works quoted in Section 2 above), we know that (1) may represent a policy function resulting from a problem of optimal growth or optimal consumption. Similar maps may describe the time evolution of consumption in a model of overlapping gnerations (see, for example, Benhabib and Day (1981) and (1982); Grandmont (1985)). Alternatively, equation (1) could describe the dynamics of inventory in a Keynesian model (see, for example, Lorenz, 1989).

Since for our present purpose the specific form of f is relatively unimportant, we select the celebrated "logistic" map, namely we put

$$f(x) = rx(1-x) \tag{2}$$

Equation (2) is known to have chaotic behaviour for many (actually for infinitely many) values of the parameter r. However, the only exact value of r for which we can prove this claim rigorously is 4. The reason for this result is that - when $r = 4$ - map (2) can be trasformed into a even simpler equivalent map. Introducing the invertible change of co-ordinates

$$y = \left(\frac{2}{\pi}\right) \arcsin \sqrt{x}, \qquad y \in [0, 1]$$

map (2) is tranformed into

$$T(y) = \begin{cases} 2y & \text{for } 0 \le y \le 1/2 \\ 2 - 2y & \text{for } 1/2 \le y \le 1 \end{cases}$$

A formidable result originally due to Ulam and von Neumann shows that this transformation preserves most dynamical properties of the map. In particular, it preserves certain properties that characterizes chaotic behaviour such as the Lyapunov characteristic exponent and metric entropy[4].

Since map (4) is semilinear and therefore easy to study, we shall concentrate our discussion on it, knowing that the basic results of the investigation can be referred back to the more "realistic" map (2).

In our previous intuitive discussion of chaos, we came to the conclusion that from the economic point of view, the fundamental characterization of chaotic dynamics is the lack of predictability, even when the rule that governs the dynamics is known exactly - i.e., when we assume that the theoretical model is the "true" model.

In order to provide a more rigorous discussion of this point, we must first of all clarify a preliminary question. If a dynamical model – a map or a system of ordinary differential equation – is known *and the state of the system at each instant of time can be measured with absolute precision*, there is nothing left to discuss. Well-known mathematical results guarantee that, under very general conditions, the future time evolution of the system is also known exactly.

Absolute precision of measurement, however, is a mathematical expression without empirical content. To be convinced of that, consider the following proposition: "The length of this rod is π centimeters." A moment's reflection will suggest that there is no way to prove that the proposition is true. All we can prove is a different proposition such as: "The length of this rod is between 3 and 4 cm. (or between 3.1 and 3.2 cm., or between 3.14 and 3.15", and so on and so forth)", where points on the real line are replaced by *intervals* which are the smaller the greater is the (finite) precision of our measuring device.

Although in economic models we routinely assume that the relevant quantities are represented by *real* numbers (i.e., numbers typically identified by an infinite number of decimals), the quantities about which real

---

[4]For a discussion of this point, see, for example, Ruelle, 1989, pp. 40-43.

economic agents make decisions must be approximations, i.e., in the case under discussion, intervals of the line.

The assumption of absolute precision of measurement in dynamical models is innocuous for simple systems, e.g. for systems whose asymptotic behaviours are described by fixed (equilibrium) points, or by periodic orbits. For these systems, introducing observational errors into the model does not change the results qualitatively. Most importantly, for simple systems *knowledge of the dynamical equations and of the past history of the system guarantee predictability of the arbitrarily distant future within the limits of precision of the mesuring device.*

The situation is entirely different if we consider chaotic systems. For these systems the behaviour of "perfect measurement dynamics" is totally different from that of "imperfect measurement dynamics". Knowledge of the "true" model and (finite precision) knowledge of the infinitely remote past is not sufficient to predict the future. Moreover, the quality of prediction is not fixed by the distorting effect of the measuring devise but deteriorates as the time distance over which prediction is performed increases. (Of course, if the dynamics of the system is bounded, the size of the error is also bounded.)

The notion of imperfect measurement may be made clearer by looking at the specific example of the "tent" map. Consider, first of all, that this map has two non-negative equilibrium points both unstable and that the interval $I = [0, 1]$ is invariant with respect to T. In what follows, we shall study the orbit of T on I.

Finite precision of observation means that at each instant of time we can locate the position of the system not on a point but on an sub-interval of I. A finite collection of N semi-open and disjoint subintervals $(I_1, \ldots, I_N)$ of I whose union is equal to I is called a *partition* of I and we shall denote it by $\mathcal{P}_0$. A partition can be also viewed as an "observation function" $\mathcal{P}_0 : I \to \{I_1, \ldots, I_N\}$, such that for any exact position of the system $x \in I \subset \mathbf{R}$, $\mathcal{P}_0(x)$ denotes the element of the partition in which x is contained.

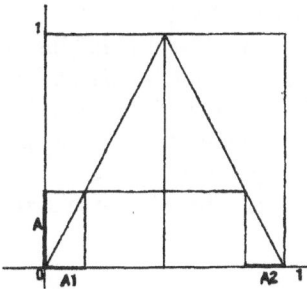

Figure 1. The Lebesgue measure m is preserved by the tent map T. Since $T^{-1}(A) = A1 \cup A2$ and $m(A1) = m(A2) = \frac{1}{2}m(A)$, $m(T^{-1}(A)) = m(A)$.

4 Chaos and Predictability. Metric Entropy

We shall now show that the dynamics of T on the partitioned space $\{I_1, \ldots, I_N\}$ is unpredictable in an appropriately defined sense and, moreover, that it may be undistinguishable from a stochastic process.

There exist of course many ways of partitioning the interval I but in this case, as we shall see in a moment, the choice of the partition does not matter. Let us then begin with the assumption that our measuring device is very rough and can only distinguish from points situated on the left or on the right of the middle point $x = 1/2$. Accordingly, we start from the crudest of all partitions of the $[0, 1]$ interval, namely $I_1 = \{0 \leq x < 1/2\}; I_2 = \{1/2 \leq x \leq 1\}$. For short, let us indicate I_1 by L (Left of the middle point) I_2 by R (Right of the middle point)[5]. If we now start from an initial point situated on L or R and iterate the map T, we shall generate a sequence of two symbols, something looking like

$$LLRLRRRL...$$

Our final goal is to verify the (un)predictability of this (partitioned) dynamics. To do this, we must preliminarly change our perspective a take a probabilistic point of view. Let us consider a single iteration of the map on the partition $\mathcal{P}$, as an experiment whose outcome is uncertain (it could be L or R) and let us try to evaluate the amount of uncertainty concerning the outcome (before the experiment), or, equivalently, the amount of information provided by the experiment (after it). We can do this by means of a formula known as "Shannon entropy" after the name of one of the founders of the modern theory of information. If ξ is a random variable taking a finite number N of values with probabilities $p_1, \ldots, p_N$, then the entropy of ξ is

$$H(\xi) = -\sum_{i=1}^{N} p_i \log(p_i) \tag{5}$$

with the convention that $0 \log 0 = 0$.

The choice of (5) can be justified by making recourse to certain axioms of probability theory which cannot be discussed here[6]. Suffice it to notice that H is maximum when we have $p_i = 1/N$, $\forall i$, i.e., the probability is uniformly distributed among the different possible outcomes and it is minimum when there is a j for which $p_j = 1$ and $p_i = 0$, for $i \neq j$, i.e. when the outcome is certain. This of course agrees with our commonsense notion of uncertainty.

---

[5] In the present context, whether the middle point $x = 1/2$ belongs to L or R does not matter.

[6] These are the so-called "Khinchin axioms", see Khinchin (1957).

To apply this notion to our system, we need to find a measure of probability μ that assigns (nonnegative) probability values to subsets of I (i.e., that tells us how likely it is that iterates of the map "visit" that subset.) The measure must "agree" with the outcome-generating map T in the sense that subsets of I of a given measure must be mapped to subsets of equal measure. Mathematically, this property is denoted by the equation

$$\mu(T^{-1}(I_i)) = \mu(I_i) \tag{6}$$

where I_i is a subset of I and $T^{-1}(I_i)$ denotes the *pre-image* of the set I_i under T[7]. In this case, we say that μ is *preserved* by T or, equivalently, that μ is *invariant* with respect to T.

The problem of finding an invariant measure (or to select among them when there are many) is a difficult one and we shall discuss it here. In our case, however, one can promptly verify that the map T preserves the so-called *Lebesgue measure*. This a mathematically sophisticated version of the common notion of length. If we take the interval $[0, 1]$ as a unit of measure of length, the Lebesgue measure (henceforth denoted by m) is the measure that assigns to subintervals values equal to their lengths. Figure 1 clearly shows that m is preserved by T.

We conclude that the information obtained by making one observation of the system on the partition $\mathcal{P} = \{L, R\}$ (alternatively: by means of the observation function $\mathcal{P} : I \rightarrow \{L, R\}$) is the following:

$$\begin{aligned} H(\mathcal{P}_l, m) &= -[m(L)\log(m(L)) + m(R)\log(m(R))] \\ &= -[1/2\log(1/2) + 1/\log(1/2)] = \log 2 \end{aligned}$$

(remember that $L = \{0 \le x \le 1/2\}$ and $R = \{1/2 < x \le 1\}$)

When dealing with the predictability of a dynamical system, we are not mainly interested in the entropy of a partition of the state space, i.e., the information contained in a single observation, but we wish to know the entropy of the system, defined as the rate at which replications of the experiments - i.e., repeated, finite-precision observations of the system as it evolves in time - produce information, when the number of observations becomes very large.

For this purpose, let us then consider the partition formed by taking the pre-image of $\mathcal{P}_0$ under T, namely $T^{-1}(\mathcal{P}_0)$ and then performs the operation

$$\mathcal{P}_1 \equiv \mathcal{P}_0 \bigvee T^{-1}(\mathcal{P}_0)$$

This operation consists in taking all the possible intersections of the elements of $\mathcal{P}_0$ and $T^{-1}(\mathcal{P}_0)$ and is called a "span", or a "joint". A moment's

---

[7]In equation (6), we make use of the pre-image of the map T to take into account the case in which T is non-invertible. If T is invertible, (6) can be equivalently written as $\mu(T(I_i)) = \mu(I_i)$.

reflection will suggest that a cell of $\mathcal{P}_1$ corresponds to a sequence of two states of the system evolving under T, observed in the partitioned state. In our case, it corresponds to a sequence such as LL, LR, RL, or RR. Shannon's entropy can again be used to measure the gain of information obtained by such (double) observation of the system - or, equivalently by performing the iteration on the map T twice.

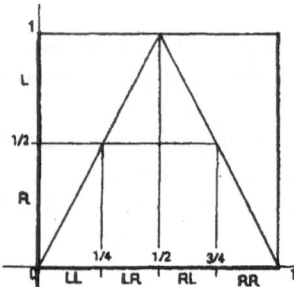

Figure 2. The partition $\mathcal{P}_0 = \{L, R\}$ and its counter-image.

Observing Figure 2, we can promptly see that the pre-image of L under T consists of the two sub-intervals $LL \equiv \{0 \le x \le 1/4\}$ and $RR \equiv \{3/4 \le x \le 1\}$, whereas the pre-image of R under T consists of the two sub-intervals $LR \equiv \{1/4 < x \le 1/2\}$ and $RL \equiv \{1/2 < x \le 3/4\}$. Then we have

$$T^{-1}(L) \cap L = LL \qquad T^{-1}(L) \cap R = RR$$

and

$$T^{-1}(R) \cap L = LR \qquad T^{-1}(R) \cap R = RL$$

It follows that the partition $\mathcal{P}_1$ consists of the four cells LL, LR, RL, RR each of (normalized) length $1/4$ and therefore of Lebesgue measure $1/4$. Consequently, we have

$$H(\mathcal{P}_1, m) = -4[1/4 \log(1/4)] = \log 4 = 2 \log 2$$

It is now easy to generalize our calculations to an arbitrary number of observations of the system, constructing the partition

$$\bigvee_{i=0}^{n} T^{-i}(\mathcal{P}_0) \tag{7}$$

By replicating the procedure just described, we can verify that partition (7) consists of 2^n cells, each of them corresponding to a sub-interval of length

2^{-n} and therefore of Lebesgue measure 2^{-n}. Thus the Shannon entropy of partition (7) is equal to

$$-2^n(2^{-n}\log(2^{-n})) = n\log 2$$

Finally let us know consider the limit for $n \to \infty$ (i.e., when the number of oservations becomes arbitrarily large) of the *average* uncertainty (=the average gain in performing just another observation). We shall have

$$h(\mathcal{P}_0, m) = \lim_{n\to\infty} \frac{1}{n} H(\bigvee_{i=0}^{n} T^{-i}\mathcal{P}_0) = \lim_{n\to\infty} \frac{1}{n}[-2^n)2^{-n}\log(2^{-n})] = \log 2 \tag{8}$$

For any given map, the quantity defined by equation (8) depends on the invariant measure m *and* on the partition $\mathcal{P}_0$. Let us now consider all possible partitions of our state space, evaluate the entropy of each of them, and find the largest one. We shall then write

$$h(m) = \sup_{\mathcal{P}} h(\mathcal{P}, m) \tag{9}$$

The quantity (9) is called "metric" or "Kolmogorov-Sinai (KS) entropy". In the sequel, the term "entropy" without qualifications will refer to the KS-entropy. Clearly when $h = 0$ we can conclude that, if we observe the (approximate) state of the system for a sufficiently long time, then there will be no uncertainty left, i.e., there will be no additional information to obtain from an extra observation. In other words, the past of the system determines its future, i.e., the sytem is predictable. On the contrary, if the entropy of the system is positive, we conclude that the knowledge of the governing map and of the infinitely remote past history of the system is not sufficient to predict its future; the system is unpredictable.

One might reasonably observe that the actual verification of equation (9) might never be possibile: how could one evaluate all possible partitions of the state space? Fortunately, there exists a result due to Kolmogorov and Sinai which guarantees that the value of the entropy calculated for a certain partition, called "generating partitions" is automatically the supreme value over all possible partitions. Given a dynamical system defined by a map T acting on a space M and a partition $\mathcal{P} = \{P_1, P_2, \ldots, P_N\}$ of M, the partition $\mathcal{P}$ is said to be *generating* if an infinite sequence of symbols P_i, $(i = 1, 2, \ldots, N)$ uniquely identifies the initial point x_0 of an orbit of the system.

From our discussion of metric entropy, it is easy to see that this is indeed the case for the tent map and the $\{L, R\}$ partition. For this map it is also true the inverse: to a given initial point x_0, there corresponds a unique infinite sequence of L, R symbols.

Inspection of equation (8) will help us understand why zero (positve) metric entropy implies predictability (unpredictability). The equation can be looked at as the limit of a fraction whose denominator is the number of observations, n which of course goes to infinity linearly with n. The numerator of (8) is the Shannon entropy of the "super-partition" $\bigvee_{i=0}^{n} T^{-1}\mathcal{P}_0$, namely:

$$H(\bigvee_{i=0}^{n} T^{-1}\mathcal{P}_0) \tag{10}$$

The cells of that "super-partition" can be interpreted as sequences of the symbols L and R of length n. Its entropy (i.e., its uncertainty) depends on two factors: (i) the number of cells, namely the number of possible n-orbits[8] of the system on the partitioned space L, R; and (ii) the probability (with respect to the invariant measure m) of each n-orbit.

It is clear that, for systems characterized by simple dynamics (e.g., orbits tending to a fixed point, or to a periodic solution), once transients have died out, the number of orbits with non-zero probability does not grow at all, and entropy must be zero. Broadly speaking this means that observers of the system will have no "surprises" and their prediction of the future course of the system will be correct (within the prescribed approximation).

It is also clear from the definition of Shannon entropy that the limit of the fraction (8) will be zero whenever the number of possible n-orbits increases less than exponentially. This happens, for example, in the often misunderstood case of quasi-periodic dynamics. In this case, the number of possible (non-zero probability) sequences is not constant but grows at a less than exponential rate. Consequently, their metric entropy is zero, uncertainty disappears asymptotically and prediction is possible.

In chaotic systems, on the contrary, entropy (8) is positive since the number of possible orbits $N(n)$ grows exponentially. This can be easily seen in the simple case of the "tent" map. For this purpose, consider that, in this case the probability of each n-orbit (i.e., the probability of each cell of partition $\bigvee_{i=0}^{n} T^{-1}\mathcal{P}_0$) is the same and equal $\frac{1}{N(n)}$. The Shannon entropy of the partition is consequently equal to

$$N(n)\frac{\log N(n)}{N(n)} = \log N(n)$$

Therefore, the formula for the metric entropy is the following

$$\frac{\log(N(n)}{n} \tag{11}$$

Clearly equation (11) tends to zero with n, unless $N(n)$ grows exponentially (or more than exponentially). We showed before that for the tent map this is indeed the case.

Heuristically speaking, the chaotic, positive-entropy case can be described as follows. Consider an observer of the system looking at a sequence

---

[8]n-orbits of course means "orbits consisting of sequences of n terms".

of n, finite precision states of a dynamical system and trying to evaluate its next state. Let each possible state be labelled by a symbol (such as L and R). Broadly speaking, uncertainty depends on the possibility of the system generating new combinations of symbols but, if the number of the possible combinations is finite, the probability of their occurrence becomes the smaller the longer the sequence and it goes to zero with $1/n$. Only if the system has the capability of generating new sequences ("surprises") at a sufficiently high rate, can uncertainty remains forever. This is precisely what happens for the "tent" map (as well as for other chaotic maps such as the "logistic").

Notice that the capability of generating "surprises" is traditionally attributed to stochastic processes. The most intriguing feature of chaotic systems is precisely that, whenever observation is not infinitely precise, they can behave stochastically. There is an important difference, however. The statistical properties of stochastic processes are typically given exogenously with respect to the deterministic (theoretical) model of the phenomena under investigation. On the contrary, the statistical properties of chaotic dynamics depend fundamentally on the structure of the model representing those phenomena.

5   Stochastic Processes and Deterministic Chaos: Bernoulli Systems

But this is not all. In fact, not all unpredictable (positive entropy) systems are equally unpredictable. Whereas zero entropy implies that the dynamics of a system is predictable with regards to any possible finite partition, positive entropy simply means that the system is unpredictable with regards to *at least one* partition. As we shall see in a moment, however, there exist systems which are totally unpredictable in the sense that they are unpredictable for *any* possible partition. Among these systems, furthermore, there exists a special class called **Bernoulli**, which is the "most chaotic", or the "least predictable" of all.

Bernoulli deterministic systems are fundamental in at least two ways. First of all, they are the essence of chaotic systems in the sense that for any chaotic (positive entropy) system, there is at least one partition of the state space on which its dynamics is isomorphic to a Bernoulli shift. This property is sometimes described by saying that all chaotic (positve entropy) systems must have a Bernoulli "factor". On the other hand, all factors of a Bernoulli system are Bernoulli.

Secondly, the output of deterministic Bernoulli systems cannot be distinguished from that of certain stochastic processes. We shall return to this point in a moment.

Keeping to the general tenor of this paper, we shall not discuss these issues in their most general and abstract way. We shall instead try to convey the basic idea of a "Bernoulli system" by means of the simplest example. In so doing, we shall show that the "tent" (and the logistic map) are indeed Bernoulli.

Let us consider again the partitioned state space Σ^2 consisting of just two elements L and R. Let us now construct an auxiliary state space whose points consist of sequences of those symbols. We can now define a map σ on Σ^2 in the following way: take a given, bilaterally infinite sequence of the two symbols L, R and shift it one step to the left, thus:

(i) Before the shift:

$$\ldots LRLLLRRLL.RRRLRLRRR\ldots$$

(ii) After the shift:

$$\ldots RLLLRRLLR.RRLRLRRRL\ldots$$

(where the arbitrary "decimal point" is introduced only for the sake of presentation). The map σ acting on the space Σ^2 in the manner indicated above is called "shift map".

Suppose now that the probability assigned to the elements L and R of the L, R space is, say, $p_L = p_R = 1/2$ and consider the measure μ_σ that assigns to each finite sequence of L, R a probability equal to the product of the probability of each element of the sequence. Clearly, μ_σ is invariant with respect to σ. We now have a dynamical system $(\Sigma^2, \sigma, \mu_\sigma)$, defined by a state space, a map and a probability measure invariant with respect to that map.

Analogous definitions are used for one-sided sequences. In this case, the space $\hat{\Sigma}^2$ is the space of (one-sided) sequences of L, R and the map $\hat{\sigma}$ acts by shifting the sequence one step to the left and dropping the first element of the sequence. In this case, the (one-sided) "shift map" will act as follows.

(i) Before the shift:
$$.RRRLRLRRR\ldots$$

(ii) After the shift:

$$.RRLRLRRR(L\,or\,R)\ldots$$

The system $(\Sigma^2, \sigma, \mu_\sigma)$ or $((\hat{\Sigma}^2, \hat{\sigma}, \hat{\mu}_\sigma,)$ is the simplest example of a Bernoulli system.

Notice that, if we rename our symbols "Head" and "Tail", the system just defined can be taken as a mathematical representation of the independent stochastic process consisting of repeated tosses of a (fair) coin.

It can be shown that the "tent" (and thereby the logistic map) is *isomorphic* to the system $(\hat{\Sigma}^2, \hat{\sigma}, \hat{\mu}_\sigma)$, where the measure $\hat{\mu}_\sigma$ is the "product measure" defined above. Isomorphism is a type of equivalence that - in the case under investigation - can be defined by the following properties:

(i) the following diagram "commutes"

$$\begin{array}{ccc} [0,1] & \xrightarrow{\quad T \quad} & [0,1] \\ h \downarrow & & \downarrow h \\ \hat{\Sigma}^2 & \xrightarrow{\quad \hat{\sigma} \quad} & \hat{\Sigma}^2 \end{array}$$

where 'commutes' means that $h \circ T = \hat{\sigma} \circ h$; T is the "tent" map; h is the invertible function mapping the set $[0,1]$ onto the space $\hat{\Sigma}^2$ of (one-sided) infinite sequences of symbols L, R as defined above; $\hat{\sigma}$ is the (one-sided) shift map.

(ii) the map h preserves the probability structure, i.e., if A and B are, respectively, measurable subsets of $[0,1]$ and $\hat{\Sigma}^2$, then $m(A) = \mu_\sigma \circ h(A)$ and $\mu_\sigma(B) = m \circ h^{-1}(B)$. Here m is the Lebesgue measure preserved by T and μ_σ is the product measure preserved by $\hat{\sigma}$ as discussed above. The commuting property (i) is easily verified by making use of the fact that the partition L, R is a generating partition for the tent map. Property (ii) can also be ascertained verifying that the probability assigned by the Lebesgue measure to a given sub-interval of $[0,1]$, i.e., its length, is equal to the probability of the set of (one-sided) sequences of the L, R symbols corresponding to the set, evaluated by means of the $\hat{\sigma}$-invariant measure μ_σ (i.e., the measure that assigns to any sequence of n symbols L, R the product of the probability of each symbol (in this case $1/2$ for both L and R).

Isomorphism preserves certain statistical properties, in particular the Lyapunov characteristic exponent and the metric entropy. The class of processes which are isomorphic to a Bernoulli process are called **B-processes**. It includes both deterministic processes - e.g., besides the the tent map, the logistic map (for $r = 4$); the Lorenz geometric model; the Šilnikov model - and stochastic processes - e.g., the i.i.d. processes; the continuous-time Markov processes. The B-processes are characterized by an asymptotic form of independence called "Very Weak Bernoulli". The latter broadly speaking implies that the future of a process conditioned on its past uniformly converges, in a certain appropriate metric, to the unconditioned past. For Bernoulli maps, the uncertainty concerning "the future" (as represented by the partition $\mathcal{P}$), conditional on "the past" (the partition $\bigvee_{i=1}^{n} T^{-i}\mathcal{P}$) asymptotically and uniformly tends to that of the unconditioned future.

The stochastic properties of the behaviour of chaotic deterministic is dramatically revealed by some recent results on B-processes. In particular, it was showed that for a deterministic system belonging to the class of B-processes we can define a stochastic k-state Markov chain such that the dynamics of the former is arbitrarily close – in an appropriately define sense – to that of the latter.

Since the mathematics behind these results is rather hard, we cannot provide a full discussion of this question here. In line with the strategy of this paper, we shall instead illustrate the main idea by means of a simple application to the tent map, referring the reader to the relevant literature (see Ornstein and Weiss, 1991; Radunskaya, 1992) for a more rigorous and general discussion and proofs.

For the tent map, and for a given partition of the state space $I = [0, 1]$, we can define a Markov chain on k states (the number of states depending on the partition), such that its sample paths are indistinguishable (within the prescribed resolution, or precision of observation) from the orbits of the deterministic tent map[9]. For example, if the partition is $\{L, R\}$, i.e., if we can only tell whether the state of the system is on the left or on the right of the middle point of the state space $[0, 1]$, then a Markov chain on two states L and R, with transition matrix

$$\begin{pmatrix} \frac{1}{2} & \frac{1}{2} \\ \frac{1}{2} & \frac{1}{2} \end{pmatrix} \tag{12}$$

will generate sample paths indistinguishable (within the prescribed degree of accuracy) from those of the tent map on the partitioned space $\{L, R\}$.

If we now increase the precision of observation so that we can now locate point in the state space within sub-intervals of length 2^{-k}), we can construct a 2^k state Markov chain with transition matrix:

$$\begin{pmatrix} 1/2 & 1/2 & 0 & 0 & \cdots & \cdots & 0 & 0 \\ 0 & 0 & 1/2 & 1/2 & \cdots & \cdots & 0 & 0 \\ \cdots & \cdots & \cdots & \cdots & \cdots & \cdots & \cdots & \cdots \\ 0 & 0 & 0 & 0 & \cdots & \cdots & 1/2 & 1/2 \\ \cdots & \cdots & \cdots & \cdots & \cdots & \cdots & \cdots & \cdots \\ 0 & 0 & 1/2 & 1/2 & \cdots & \cdots & 0 & 0 \\ 1/2 & 1/2 & 0 & 0 & \cdots & \cdots & 0 & 0 \end{pmatrix} \tag{13}$$

Again it will not be possible (within the prescribed precision) to distinguish between a typical sample path of the Markov chain and an orbit of the tent map on the 2^k-partitioned state space.

The same exercise could be performed in relation to the "logistic" map with chaotic parameter, although in that case the construction of the transition matrix would be more difficult.

6 Conclusion

The implications for economics of the results just obtained are puzzling. For example, consider the case in which a model of optimal growth gives

---

[9]I owe this example to a private correspondence with Amy Radunskaya.

rise to a dynamic equation of the "logistic" type with chaotic parameter. The sequences thus generated are optimal in the sense that they solve a problem of intertemporal maximization of rational agents, in an economy satisfying the requirements of competitive equilibrium at each point of time. In the absence of exogenous, random disturbances, along optimal trajectories agents' expectations are supposed to be always fulfilled, i.e., we assume agents' perfect foresight.

While the latter assumption may be acceptable when the dynamics of the system are simple (e.g., convergence to a steady state or to a periodic orbit), it makes little sense if the dynamics are chaotic, eminently so if they are Bernoulli. In our case, the information set on which agents base their predictions typically consists of the observations of past values of the relevant variables. But we have just demonstrated that - if the entropy of the system is positive - knowledge of the infinitely remote past with arbitrarily (but not infinitely) great precision of measurement is not sufficient to forecast future values correctly. Moreover, when the system is Bernoulli, agents face sequences of values which have the same probabilistic structure as random processes. Intuitively speaking, in this case the past no longer determines the future and the assumption of perfect foresight lacks any motivation.

The equivalence result between (Bernoulli) deterministic and (Markov) stochastic processes discussed above can be given a more or less optimistic interpretation according to one's point of view and temperament. The results discussed above indicate that we cannot hope of providing a generally valid test for distinguishing deterministic chaos and "true" randomness. This would certainly be impossible for Bernoulli systems and, to the extent that the conjecture that "most observable chaos is Bernoulli" (see Ornstein and Weiss *op. cit*, p. 22.) is correct, it would be generally impossible. Consequently, at least for a certain class of concrete dynamical systems, the possibility exists of representing them either as deterministic systems (plus perhaps some random disturbances), or as stochastic processes. The choice between the two is a matter of expedience rather than theory. In principle, a deterministic representation is superior for explanation purposes, but this is only true if we can provide a physical (economic) interpretation of the state variables and the functional relationships among them.

References

[1] Benhabib, J. and Day, R.H. (1981). Rational choice and erratic behaviour. *Review of Economic Studies*, **48**, 459–471.

[2] Benhabib, J. and Day, R.H. (1982). A characterization of erratic dynamics in the overlapping generations models. *Journal of Economic*

Dynamics and Control, **4**, 37–55.

[3] Blanchard, O.J. and Fischer, S. (1987). *Lectures on Mathematics.* Cambridge, Mass.: MIT Press.

[4] Boldrin, M. and Montrucchio, L. (1986). On the indeterminacy of capital accumulation paths. *Journal of Economic Theory*, **40**, 26–39.

[5] Deneckere, R. and Pelikan, S. (1986). Competitive chaos. *Journal of Economic Theory*, **40**, 13–25.

[6] Farmer, J.D. and Sidorowich, J.J. (1987). Predicting chaotic time series. *Physics Review Letters*, **59**, 845–848.

[7] Ford, J. (1983). How random is a coin toss? *Physics Today*, **36**, 40–48.

[8] Frisch, R. (1933). Propagation problems and impulse problems in dynamic economics. In *Economic Essays in Honour of Gustav Cassel*, 171–205. London: Allen and Unwin.

[9] Goodwin, R.M. (1951). The nonlinear accelerator and the persistence of business cycles. *Econometrica*, **19**, 1–17.

[10] Grandmont, J.M. (1985). On endogenous competitive business cycles. *Econometrica*, **53**, 995–1045.

[11] Granger, C.W.J. and Newbold, P. (1977). *Forecasting Economic Time Series.* New York: Academic Press.

[12] Hicks, J.R. (1950). *A Contribution to the Theory of the Trade Cycle.* Oxford: Oxford University Press.

[13] Kaldor, N. (1940). A model of the trade cycle. *Economic Journal*, **50**, 78–92.

[14] Khinchin, A.I. (1957). *Mathematical Foundations of Information Theory.* New York, Dover Publ.

[15] Lorenz, H.W. (1989). *Nonlinear Dynamical Economics and Chaotic Motion.* New York: Springer Verlag.

[16] Mañé, R., 1987, *Ergodic Theory and Differentiable Dynamics*, (Springer-Verlag, Berlin)

[17] Medio, A., 1992, *Chaotic Dynamics. Theory and Applications to Economics*, Cambridge University Press

[18] Medio, A. and G. Negroni, 1996, Chaotic Dynamics in Overlapping Generations Models with Production, in: Barnett, W.A., A. Kirman and M. Salmon (eds.), Nonlinear Dynamics and Economics, Cambridge University Press

[19] Ornstein, D.S. and B. Weiss, 1991, Statistical Properties of Chaotic Systems, *Bulletin (New Series) of the American Mathematical Society* **24**, N.1, 11-115.

[20] Radunskaya, A., 1992, *Alpha-Congruence of Bernoulli Flows and Markov Processes: Distinguishing Random and Deterministic Chaos*, unpublished Ph.D. Thesis (Stanford University).

[21] Ruelle D., 1989, *Chaotic Evolution and Strange Attractors*, (Cambridge University Press, Cambridge).

[22] Slutsky, E. (1927). The summation of random causes as the source of cyclical processes. *Econometrica*, **5**, 105–146.

[23] Yule, G.U. (1927). On a method of investigating periodicities in disturbed series. *Philosophical Transactions*, **226A**, 267–298.

Long-Term Memory and Chaos: a Note

Gilbert ABRAHAM-FROIS, Sandrine LARDIC and Valérie MIGNON

1 Introduction

Is it possible to study jointly long-term memory processes and chaotic processes? In a first sight, this interrogation can appear somewhat surprising because of the properties showed by the two kinds of processes. Effectively, a long-term memory process, like an ARFIMA process, is a stochastic one, while a chaotic process is by definition a deterministic one. However, this question finds its origins in recent works of Peters (1991, 1994) setting number of relations between the two processes. On the one hand, Peters showed that long-term memory and chaos concepts can be linked by the mean of the fractal dimension: "*Fractal time series are characterized as long memory processes. They possess cycles and trends, and are the result of a nonlinear dynamic system, or deterministic chaos*" (Peters, 1991, p. 119). In this way, a long-term memory process would be a process whose attractor has a fractal dimension. We can thus think that the link between fractal dimension and chaos might be set by the mean of the concept of strange attractor. However, we will show that this relation is useless in order to detect the presence of chaotic dynamics. On the other hand, according to Peters, long-term memory can be equivalently detected by R/S analysis and Lyapunov exponents since there exist nonperiodic cycles for the two processes: "*The long memory effect in equity prices has now been confirmed by two separate types of nonlinear analysis. R/S analysis on monthly S&P 500 stock returns found a biased random walk with a memory length about four years. The Lyapunov exponent for monthly inflation-detrended S&P 500 prices found a 42-month cycle*" (Peters, 1991, p. 180).

This note proposes a critical study of links established by Peters (1991, 1994) between chaotic and long-term memory processes. To this end, after having presented definitions of the main concepts concerning long-term memory and chaos, we study the relations between the two notions. We show that these links are only superficial and erroneous.

2 Long-term memory and chaotic processes: definitions

In a first time, we define the main long-term memory processes. The first long-term memory processes have been presented in terms of fractal processes because of their self-affinity property (fractional Brownian motion and fractional Gaussian noise). In a second time, we define the main concepts relative to chaotic dynamics.

2.1 LONG-TERM MEMORY REPRESENTATION: FRACTAL PROCESSES AND ARFIMA PROCESSES

The first long-term memory process, the so-called fractional Brownian motion, was defined by Mandelbrot and Van Ness (1968) and constitutes a generalization of the ordinary Brownian motion for which increments were independent. H, the key parameter of this process, characterizes the self-affinity property of fractional Brownian motion. This parameter is called the Hurst exponent in reference to the English hydrologist H.E. Hurst. The Hurst exponent is particularly interesting because it allows to classify time series according to their dependence structure: null or short-term memory, long-term memory (Joseph effect) and anti-persistence phenomenon. Because economic series are discrete time series, Mandelbrot and Wallis (1969) have derived a discrete time analogue — the fractional Gaussian noise — of the continuous time fractional Brownian motion. This process also exhibits the persistence phenomenon. Another class of discrete time models linked to fractional Brownian motion is constituted by ARFIMA (Auto-Regressive Fractionally Integrated Moving Average) processes. These models are a generalization of standard ARIMA(p, d, q) processes for which the differencing parameter d was an integer. Traditionally, econometricians consider either non stationary series (*i.e.* series with a unit root) or stationary series (*i.e.* series without such a root). In the first case, time series are modeled by ARIMA(p, d, q) processes, with d generally equal to one, and, in the second case, by ARMA(p, q) processes (*i.e.* $d = 0$). ARIMA processes refer to an infinite memory and ARMA processes to short-term memory. Hence, the intermediate case of long-term memory, which corresponds to a fractional d, has been ignored. In order to take into account the persistence phenomenon, Granger and Joyeux (1980) and Hosking (1981) have developed ARFIMA(p, d, q) processes where d can take real values and not only integer ones. These processes are linked to the Hurst exponent by the mean of a relationship between the fractional differencing parameter and the Hurst exponent, derived from the autocovariance function of fractional Gaussian noise.

Generalization of Brownian motion: fractional Brownian motion

A Brownian motion $B(t, \omega)$, is a continuous-time stochastic process given by:

$$B(t, \omega) = \int_{-\infty}^{t} W(s) \, ds$$

where $W(s)$ is a Gaussian white noise.

A Brownian motion is characterized by independent Gaussian increments and a spectral density of ω^{-2}. In other words, $B(t, \omega)$ is such that:

- $B(t_2, \omega) - B(t_1, \omega)$ has zero mean and a variance proportional to $|t_2 - t_1|$.
- $B(t_2, \omega) - B(t_1, \omega)$ and $B(t_4, \omega) - B(t_3, \omega)$ are independent, if the intervals (t_1, t_2) and (t_3, t_4) do not overlap.

Standard deviation of the increment $B(t + T, \omega) - B(t, \omega)$, for $T > 0$, is equal to $T^{\frac{1}{2}}$, which is known as the $T^{\frac{1}{2}}$ law.

Because of the independence between increments, the process displays only short-term memory. Mandelbrot and Van Ness (1968) proposed an extension of this ordinary Brownian motion in order to take into account the long-term memory phenomenon, *i.e.* the possibility of a dependence between distant observations. This generalization gives the so-called fractional Brownian motion.

Fractional Brownian motion, denoted as $B_H(t, \omega)$, is a zero mean Gaussian stationary process, defined as:

$$B_H(t, \omega) - B_H(0, \omega) = \frac{1}{\Gamma(H + \frac{1}{2})} \left\{ \begin{array}{l} \int_{-\infty}^{0} \left[(t - s)^{H - \frac{1}{2}} - (-s)^{H - \frac{1}{2}} \right] dB(s, \omega) \\ + \int_{0}^{t} (t - s)^{H - \frac{1}{2}} dB(s, \omega) \end{array} \right\}$$

where $b_0 = B_H(0, \omega)$
if $b_0 = 0$, we have:

$$B_H(t, \omega) = \frac{1}{\Gamma(H + \frac{1}{2})} \left\{ \int_{0}^{t} (t - s)^{H - \frac{1}{2}} dB(s, \omega) \right\}$$

Thus, if $H = \frac{1}{2}$, fractional Brownian motion reduces to ordinary Brownian motion and becomes a short-term (or null) memory process.

To simplify notations, let us write $B_H(t)$ for $B_H(t, \omega)$. The variance of fractional Brownian motion is given by t^{2H} and its covariance:

$$E\left[B_H(s), B_H(t)\right] = \frac{1}{2} \left[s^{2H} + t^{2H} - |s - t|^{2H} \right]$$

The key parameter of fractional Brownian motion is H, where $0 < H < 1$, which is called the Hurst exponent. This parameter appears in a relationship linking $\Delta B_H(t)$ and Δt:

$$\Delta B_H(t) \sim \Delta t$$

where

$$\Delta B_H(t) = B_H(t_2) - B_H(t_1) \quad \Delta t = t_2 - t_1$$

This relationship illustrates the self-affinity property of fractional Brownian motion. In other words, this kind of process is statistically self-affine in the sense that $\{B_H(rt), t \geq 0\}$ has the same finite dimensional distribution as $\{r^H B_H(t), t \geq 0\}$ for all $r \geq 0$ [1]

From the covariance function of fractional Brownian motion, we can derive the correlation between observations separated by a very long time span as:

$$C = 2^{2H-1} - 1$$

In this sense, fractional Brownian motion is a long-term memory process. Actually, we know that, in the case of a short-term memory process, the correlation goes to zero when the distance between the two points under consideration goes to infinity. In the case of fractional Brownian motion, this correlation tends to $2^{2H-1} - 1 \neq 0$ if $H \neq \frac{1}{2}$. From this correlation, it is possible to classify time series according to their dependence structure:

• when $H = \frac{1}{2}$, fractional Brownian motion reduces to ordinary Brownian motion. The correlation C is null and the process does not exhibit long-term memory phenomenon.

• when $\frac{1}{2} < H < 1$, the correlation C is positive. The process displays long-term memory and exhibits the persistence phenomenon.

• when $0 < H < \frac{1}{2}$, the correlation C is negative. The process is anti-persistent.

Economic time series are typically discrete time series. Hence, in order to model these series by a long-term memory process, it is necessary to derive a discrete time analogue of continuous time fractional Brownian motion. This is accomplished with the discrete time fractional Gaussian noise and ARFIMA processes.

Discrete time: fractional Gaussian noise and ARFIMA processes

Definitions

Fractional Gaussian noise.

The fractional Gaussian noise Y_t, introduced by Mandelbrot and Wallis (1969), is the increment of fractional Brownian motion:

$$Y_t = B_H(t) - B_H(t - 1)$$

---

[1]Note that self-affinity is different from self-similarity although the two notions are widely assimilated in the literature. Unlike self-similar curves, a self-affine process (like the fractional Brownian motion) requires different scaling factors in the two coordinates: r for t, but r^H for $B_H(t)$ reflecting the special status of the t coordinate. Each t correspond to only one value of B_H but any specific B_H may occur at multiple t's. Such non-uniform scaling is known as self-affinity rather than self-similarity.

It is a zero mean, Gaussian stationary process whose autocovariance function is given by:

$$\gamma(k) = \frac{1}{2}\left[|k+1|^{2H} - 2|k|^{2H} + |k-1|^{2H}\right], k \geq 0$$

and

$$\lim_{k \to \infty} k^{2-2H}\gamma(k) = H(2H-1)$$

The spectral density is given by:

$$f(\omega; H) = \sigma^2(2H)^{-2H-2}\Gamma(2H+1)\sin(\pi H)4\sin^2\left(\frac{\omega}{2}\right)\sum_{n=-\infty}^{\infty}\left|n + \left(\frac{\omega}{2\pi}\right)\right|^{-2H-1}$$

and $\quad \lim_{\omega \to 0} \omega^{2H-1}f(\omega; H) = \left(\frac{2\sigma^2}{\pi}\right)\Gamma(2H+1)\sin(\pi H)$

Thus, by analogy with the fractional Brownian motion, when $H = \frac{1}{2}$, $\gamma(k) = 0$ and the Y_t's are independent. When $\frac{1}{2} < H < 1$, the Y_t's are positively correlated and they display long-term memory. Finally, when $0 < H < \frac{1}{2}$, the process exhibits anti-persistence phenomenon.

Thus, fractional Gaussian noise is a discrete time analogue of the continuous time fractional Brownian motion. However, the definition of this process is essentially based on its autocorrelation function[2]. The idea of Granger and Joyeux (1980) and Hosking (1981) is to develop a truly discrete time process by looking for a discrete time version of fractional Brownian motion rather than a covariance function. This gives rise to ARFIMA(p, d, q) processes (Auto-Regressive Fractionally Integrated Moving Average). They generalize the standard ARIMA processes in which the differencing exponent d was an integer. In the case of ARFIMA processes, d can take real values and not only integer ones. A fractionally integrated time series is characterized by a dependence between distant observations as one can observe it in the autocovariance function or in the spectral density function. One would remark thus that these types of processes allow to reduce the constraints on autoregressive and moving average coefficients of parametric models[3].

ARFIMA processes.

These processes constitute a generalization of ARIMA(p, d, q) processes for which the differencing parameter d was an integer. In the case of

---

[2]Recall that its autocorrelation function is the same as whose of the fractional Brownian motion in first differences.

[3]A long-term memory process can always be approximated by an ARMA(p, q) process, but the orders p and q necessary to obtain a relatively good approximation can be too great and render parameters estimation very difficult.

ARFIMA processes, d can take real values and not only integer ones. The case of a non-integer differencing parameter is important in terms of long-term memory. In ARMA(p, q) processes, $d = 0$, and refers to a time series which has zero (or short-term) memory. In ARIMA(p, d, q) processes, d is typically equal to one and refers to infinite memory. Thus the intermediate case, that is d fractional, is ignored. Typically, the class of ARFIMA processes can be used to model data dependence which is stronger than allowed in stationary ARMA processes and weaker than implied by unit root processes.

ARFIMA$(0, d, 0)$ process

The analogous in discrete time of Brownian motion is the random walk or ARIMA$(0, 1, 0)$:

$$(1 - L)X_t = u_t$$

where u_t is an independently and identically distributed (iid) variable. Thus the first difference of X_t is the white noise u_t in discrete time. By analogy, one defines the fractional white noise of parameter d by:

$$(1 - L)^d X_t = u_t$$

where u_t is a white noise and $(1 - L)^d$ is given by the binomial expansion:

$$\nabla^d = (1 - L)^d = 1 - dL - \frac{d(1 - d)}{2!}L^2 - \frac{d(1 - d)(2 - d)}{3!}L^3 - \ldots = \sum_{j=0}^{\infty} \pi_j L^j$$

where

$$\pi_j = \frac{\Gamma(j - d)}{\Gamma(j + 1)\Gamma(-d)} = \prod_{0 < k \leq j} \left(\frac{k - 1 - d}{k} \right) \qquad j = 0, 1, \ldots$$

and

$$\Gamma(x) = \begin{array}{ll} \int_0^{\infty} t^{x-1} e^{-t} dt & \text{if } x > 0 \\ \infty & \text{if } x = 0 \\ x^{-1}\Gamma(1 + x) & \text{if } x < 0 \end{array}$$

Such a defined process is called ARFIMA$(0, d, 0)$. As the properties of this model have been widely studied by Hosking (1981), we just synthesized here his main results.

Let $\{X_t, t = 0, 1, \ldots\}$ be an ARFIMA$(0, d, 0)$ process. Then:

(i) When $d < \frac{1}{2}$, $\{X_t\}$ is stationary

(ii) When $d > -\frac{1}{2}$, $\{X_t\}$ is invertible.

Hence, the $\{X_t\}$ process is stationary and invertible only if $-\frac{1}{2} < d < \frac{1}{2}$.

The autocorrelation function decreases at an hyperbolic rate, thus much slower than the simple ARMA autocorrelation function (which decreases at a geometric rate). Because of this property, ARFIMA processes are called long-term memory processes.

The autocovariance function of ARFIMA$(0, d, 0)$ process is given by:

$$\gamma(k) = \frac{\Gamma(1 - 2d)\Gamma(k + d)}{\Gamma(d)\Gamma(1 - d)\Gamma(k + 1 - d)}$$

When k goes to infinity, we have:

$$\lim_{k \to \infty} \gamma(k) = \frac{\Gamma(1 - 2d)}{\Gamma(d)\Gamma(1 - d)}|k|^{2d-1}$$

The main advantage of ARFIMA$(0, d, 0)$ process over fractional Gaussian noise is that it has a particularly simple spectral density:

$$f(\omega, d) = \left(2\sin\frac{\omega}{2}\right)^{-2d} \quad \text{for } 0 < \omega \leq \pi$$

and $\lim_{\omega \to 0} f(\omega, d) = |\omega|^{-2d}$

Generalization: ARFIMA(p, d, q) process

The ARFIMA$(0, d, 0)$ process is a special case of ARFIMA(p, d, q) process where $d \in \left]-\frac{1}{2}, \frac{1}{2}\right[$ which can be defined like this:

$$\Phi(L)X_t = \Theta(L)\epsilon_t$$

where:

$\epsilon_t = \nabla^{-d}u_t$

$u_t : WN(0, \sigma^2)$

$\Phi(L)$ and $\Theta(L)$ are lag polynomials of p and q degrees respectively.

Hence,

$$X_t - \phi_1 X_{t-1} - \ldots - \phi_p X_{t-p} = \epsilon_t + \theta_1 \epsilon_{t-1} + \ldots + \theta_q \epsilon_{t-q}$$

with

$$\epsilon_t = u_t + du_{t-1} + \frac{d(d+1)}{2!}u_{t-2} + \frac{d(d+1)(d+2)}{3!}u_{t-3} + \ldots$$

As in the case of ARFIMA$(0, d, 0)$ processes, ARFIMA(p, d, q) processes are long-term memory, stationary, and invertible when $d \in \left]-\frac{1}{2}, \frac{1}{2}\right[$ and $d \neq 0$.

Relationship between Hurst exponent and fractional differencing parameter

If we compare the autocovariance functions of fractional Gaussian noise and ARFIMA processes, we see that they have the same power decay. Thus, relating the exponents in these two expressions gives:

$$d = H - \frac{1}{2}$$

From this remarkable relationship it is possible to classify time series according to their dependence structure as in the case of the Hurst exponent:

• if $d = 0$, ARFIMA$(p, 0, q)$ process becomes a stationary ARMA process and exhibit only short-term memory. The spectral density at frequency zero is finite and positive.

• if $0 < d < \frac{1}{2}$, the process is persistent: it is a stationary process with long-term memory. Autocorrelations are positive and decay hyperbolically to zero as the lag grows. The spectral density is concentrated around low frequencies: it increases to very high values as the frequency approaches zero.

• if $-\frac{1}{2} < d < 0$, the process is anti-persistent. The spectral density is dominated by high frequency components and declines to zero as the frequency approaches zero.

Following section describes and defines some properties of chaotic processes.

Chaotic processes

Introduction: Non-linearity and determinism

A chaotic process is a non-linear deterministic process. Chaotic processes are necessarily nonlinear. It is well known that linear models can only generate four types of behavior: oscillatory and stable, nonoscillatory and stable, oscillatory and explosive, nonoscillatory and explosive. On the contrary, nonlinear models can generate much richer types of dynamics. Moreover, despite their deterministic characteristic, trajectories generated by chaotic processes appear completely random by standard linear time series methods. Deterministic chaos is characterized by self-sustained oscillations whose period and amplitude are nonrepetitive and unpredictable.

Following Brock (1986), a time series $\{y_t\}, t = 1, \ldots, T$, has a deterministic explanation if there exist a system $\{h, f, X_0\}$ such that h maps $\Re^n$ to $\Re$, f maps $\Re^n$ to $\Re^n$ and:

• $y_t = h(X_t) \forall t$
• $X_t = f(X_{t-1})$
• X_0 is the initial condition at $t = 0$.

The map f is the unknown deterministic law of motion.

The most well known example of chaotic process is given by the logistic map:

$$X_t = \mu X_t(1 - X_t), \text{ where } 0 \leq X_t \leq 1 \text{ and } 0 \leq \mu \leq 4.$$

The dynamics of this equation depends on the value of μ. If $0 \leq \mu \leq 3$ the system converges to a stationary state. However, as μ increases beyond 3, the system undergoes various qualitative changes, known as bifurcations.

For $\mu > 3$ there are cyclic solutions. The system goes through an infinite number of period-doublings: stable cycles of length 2, 4, 8, 16, ..., 2^n ,...successively appear and become unstable. This sequence continues until a limit point μ_∞ which is approximately equal to 3.57. At this point the system behaves in a very complex manner. We can illustrate some properties of this equation by plotting the bifurcation diagram[4] (see Abraham-Frois and Berrebi (1995)). For values of μ between 3.57 and 4, the bifurcation diagram exhibits black bands which correspond to chaos separated by windows. In these windows, odd-order cycles arise. For example, for $3.828\ldots < \mu < 3.849\ldots$ a stable cycle of period 3 emerges. It is successively replaced by stable cycles of period 6, 12, ..., 3.2^n , ... These bifurcations accumulate at a limit point where the behavior is similar to that at μ_∞. This is generally referred to as order within chaos: chaotic systems have an underlying order. Moreover, a further property of the logistic equation is that if any window is magnified, the diagram displays an exact copy of the complete diagram. This illustrates the self-similarity characteristic of the logistic map. For Peters (1991, 1994) this characteristic illustrates the fact that there exist a link between fractals and chaos. A further property of chaotic systems concerns the presence of an attractor.

Attractors and sensitive dependence on initial conditions

Intuitively, we can define an attractor as a subset toward which almost all sufficiently close trajectories are attracted asymptotically. Thus, an attractor gives a global picture of the long-term behavior of a dynamical system.

More formally we retain the following definition.

Definition:

Given the map f such that: $x_{t+1} = f(x_t)$

A compact set A is an attractor if there is an open set U in the neighborhood of A in phase space such that: $f^t(x) \to A$ if

$t \to \infty$ for $x \in U$. Union of all U-neighborhoods of A is the basin of attraction of A.

There exist several types of attractors. The most well-known are the fixed point, the limit cycle and the torus:

• Attracting fixed point: trajectories are attracted toward a fixed point equilibrium, which reflects a time independent solution. It is a steady state: $f^t(x_p) = x_p, \forall t$. This fixed point is the typical attractor of linear systems and constitutes the typical solution of a great number of economic models.

• Attracting limit cycle: attractor of a periodic function, which is characterized by its amplitude and period. Trajectories converge toward this closed curve, named limit cycle. Recall that a point x is a periodic

---

[4]The bifurcation diagram is the set of possible solutions in the equation.

point of period T if $f^T(x) = x$, but $f^t(x) /= x$ for $0 < t < T$. $\Phi = \{f^t(x) : 0 < t < T\}$ is the corresponding periodic orbit.

• The r-dimensional torus T^r , $r \geq 2$: this is the attractor for quasiperiodic motion with r different periods.

These three attractors are described periodically by trajectories of the considered system. The behavior of these systems is predictable: given the initial state, it is possible to predict future states. Moreover, systems which have these attractors are not sensitive upon initial conditions: two nearby trajectories will not diverge over time. But, the main characteristic of chaotic process is precisely the sensitive dependence on initial conditions whose a definition can be state as follows.

Definition:

A map f on a metric space X is sensitive on initial conditions if there is a number r, $r > 0$, such that for all $x \in X$ and for all $\epsilon > 0$ there exist:

• a point $y \in X$ such that $d(x, y) < \epsilon$
• and an integer k, $k \geq 0$, such that $d\left(f^k(x), f^k(y)\right) \geq r$

where $d(x, y)$ is the distance between x and y and f^k is the kth iterate of f.

Thus, sensitive dependence on initial conditions means that the distance between two points that were initially separated by an infinitesimal amount grows exponentially fast on the average. This is the main property of chaotic processes. This sensitivity is formalized by the condition that the largest Lyapunov exponent, which measures the rate of spread of nearby trajectories in phase space, be positive.

In order to illustrate the sensitivity on initial conditions, the concept of strange attractor was introduced by Ruelle and Takens (1971). However the term "strange" is somewhat ambiguous. It refers also to the fractal nature of attractors[5]. In many well-known examples, chaotic attractors are also strange; e.g., the Hénon map exhibits exponential divergence of neighborhood trajectories and it has a Cantor set structure. However, there exist attractors which are fractal but not chaotic: for these attractors, there is no sensitive dependence on initial conditions[6]. On the contrary, there exist attractors which are chaotic but not fractal[7] (their dimension is an integer). Thus, we define a strange attractor to be an attractor with fractal structure. We define a chaotic attractor to be an attractor with

---

[5]In general, a fractal object is an object for which the Hausdorff dimension is different from the topological dimension, and usually not an integer. Moreover, fractal objects are self-similar.

[6]See Grebogi, Ott, Pelikan and Yorke (1984).

[7]Arnold's cat attractor is chaotic, but not fractal, see Eckmann and Ruelle (1985).

sensitive dependence on initial conditions. Consequently, a system which has a chaotic attractor is said to be chaotic. A consequence of the sensitivity upon initial conditions is the long run unpredictability.

Long-term prediction.

Long-term predictions of chaotic systems are virtually impossible, because errors in measurement of the initial state propagate exponentially fast. Because we can never measure current conditions to an infinite amount of precision, we cannot predict where the process will go in the long term. Thus, the error, even if very small, will increase and affect the macroscopic state of the system. Consequently, the least error in the initial conditions render long run prediction impossible. This is another fundamental property of chaotic systems: even if these systems are deterministic, their behavior is unpredictable, except perhaps in the very short run. Moreover, this behavior can emerge in a system with a small number of degree of freedom: only one degree of freedom in discrete time (logistic map)[8] and only three in continuous time (Lorenz model). For this reason, this kind of chaos is called deterministic chaos or low dimensional chaos. The following section studies links established by Peters (1991, 1994) between long-term memory and chaos.

3 Relations between ARFIMA and chaotic processes

According to Peters (1991, 1994) long-term memory and chaotic processes can be jointly analyzed since there exist links between the two processes. On the one hand, we will see that a link between long-term memory and chaotic processes might come from a relationship between fractal dimension, Hurst exponent and fractional differencing parameter of ARFIMA processes. However, we will show that such a relation cannot be used in order to detect the presence of chaos. On the other hand, according to Peters (1991, 1994), calculation of Lyapunov exponents allows to detect the presence of long-term memory in a time series. In this sense, Lyapunov exponents might be linked with R/S analysis.

Beside these possible similarities, we will insist on the major differences between long-term memory processes and chaotic processes. On the one hand, long-term memory processes are predictable on the long run while chaotic processes are unpredictable because of sensitivity on initial conditions. On the other hand, a chaotic process is a deterministic one, while a

---

[8]More precisely, a necessary condition for a discrete system to exhibit chaotic dynamics is that it has at least one degree of freedom if it is irreversible (logistic map) and at least two degrees of freedom if it is reversible (Hénon map).

long-term memory process, such an ARFIMA process, is a stochastic one.

3.1 LONG-TERM MEMORY AND FRACTAL DIMENSION

Can we establish a link between series whose underlying process is characterized by a long-term memory and series whose underlying process has a fractal dimension? In other words, does one of these characteristic imply the second one?

In a first time, it can be tempting to conclude to the existence of such a relation between long-term memory and fractal dimension. Effectively, the Hurst exponent H is linked to fractal dimension D according to the relation $D = 2 - H$. This dimension is the dimension of the time path of fractional Brownian motion. Moreover, we know that fractional differencing parameter d of ARFIMA processes is linked to Hurst exponent by the relation $d = H - \frac{1}{2}$. Thus, the fractal dimension of time path of ARFIMA processes is given by: $D = \frac{3}{2} - d$. It is possible to establish the following classification:

- if $d \in\ ]-\frac{1}{2}; 0[$ then $D \in\ ]\frac{3}{2}; 2[$
- if $d = 0$ then $D = \frac{3}{2}$
- if $d \in\ ]0; \frac{1}{2}[$ then $D \in\ ]1; \frac{3}{2}[$.

Thus, if $H = \frac{1}{2}$ ($d = 0$), the dimension is $\frac{3}{2}$ which is effectively a fractal dimension. But $d = 0$ corresponds to a series whose underlying process is not characterized by long-term memory. As a consequence, there is no equivalence between fractal dimension and long-term memory. In other words, a time series can be characterized by an underlying process which has a fractal dimension without being a long-term memory process. This is the case of Brownian motion which is a short-term memory process characterized by a fractal dimension of $\frac{3}{2}$.

However, we observe that a long-term memory series is characterized by a fractal dimension between $\frac{3}{2}$ and 2 in case of anti-persistence and between 1 and $\frac{3}{2}$ in case of persistence. From these relations, it might be concluded that, on the one hand, a long-term memory series has a fractal dimension and, on the other hand, a series with fractal dimension is not necessarily a long-term memory series.

Nevertheless, beyond these relations between long-term memory and fractal dimension, a remark must be state: the value of fractional differencing parameter cannot allow us to conclude if a series is really characterized by a long-term memory. It must be studied together with the standard deviation of d in order to decide if d is significantly different from zero. In other words, the measure of the persistence degree in a series is fundamentally based on statistical tools: calculation of d is the result of an estimation procedure. The estimation obtained for d is a random variable. In fact, when we use various methods for estimating d, results can be sometimes divergent (e.g., Lardic and Mignon 1996a,b,c). We have previously run estimations for various series of exchange rates, stock returns and macroeconomic series which illustrate the random character of this parameter: these procedures (Geweke and Porter-Hudak, exact maximum likelihood, R/S analysis...) do not always detect long-term memory in same series, and the signs of d are sometimes opposite (table 1).

This table relates the estimation of d according to R/S analysis, modified R/S analysis (Lo, 1991), Geweke and Porter-Hudak method, and Sowell procedure (exact maximum likelihood). Real monthly data in logarithmic first differences, January 1974–November 1994. In brackets: t-value of d.

This table leads us to a second remark: is the relation between d and H empirically verified? If we accept this relation from a theoretical point of view, it seems somewhat unreasonable to use it empirically. This conclusion directly comes from our previous remark: being not able to determine a unique estimated value of d, how to determine a unique relation between d and H? Table 1 illustrates this difficulty: the use of the theoretical relation $d = H - \frac{1}{2}$ gives estimations of d which are very different from these obtained by using other methods.

Nevertheless, the dimension obtained by the relation $D = 2 - H$ is the dimension of a stochastic process, such as an ARFIMA process. Thus, it cannot be the dimension of a chaotic process (or of its attractor) which is, by definition, a deterministic process. Moreover, it should be pointed out that the dimension given by the relation $D = 2 - H$ is the dimension of a time path, *i.e.* a time series, and not the dimension of an attractor. Effectively, the dimension of a time series is between 1 et 2 whereas the dimension of an attractor can be more important. Thus, the use of the relation $D = 2 - H$ is useless if the object under study is the calculation of a dimension of a chaotic attractor. Moreover, we have previously mentioned that a chaotic attractor is not necessarily an attractor with fractal dimension. Thus, finding a fractional dimension is not enough to conclude in terms of chaotic dynamics[9].

As a consequence, if, as suggested by Peters (1991, 1994), there exist a link between long-term memory and chaotic processes, this link cannot be set through the relation $D = 2 - H$. A second relation between long memory and chaos has been established by Peters by the mean of R/S analysis and Lyapunov exponents.

3.2 LONG-TERM MEMORY AND CHAOS: R/S ANALYSIS AND LYAPUNOV EXPONENTS

We have previously observed that, because of the sensitivity on initial conditions, chaotic processes are unpredictable on the long run. On the

---

[9]See the previous distinction between strange attractor and chaotic attractor.

contrary, a long memory process is predictable on the long run. We have thus here a fundamental opposition between the two processes. Despite this fact, Peters (1991, 1994) showed that the link between long-term memory and chaos can be obtained through R/S analysis and Hurst exponent. The object of R/S analysis is to detect the presence of long-term memory in a time series. Moreover this statistic is able to detect nonperiodic cycles whose we can calculate the average length. This length corresponds to time during which an exogenous chock will affect the system. Peters (1991) found that the underlying process of American stock index SP 500 is a long-term memory one, since the estimated Hurst exponent is 0.78. The duration of this memory is about 48 months. By using tools for detecting chaos, Peters found that the value of the largest Lyapunov exponent is 0.0241 bit/month. As a consequence, if we could measure initial conditions to one bit of precision, we would lose all predictive power after 1/0.0241, or 42 months'time.Since the Lyapunov exponent is positive and the average cycle length obtained by R/S analysis is roughly equal to the length estimated using Lyapunov exponent, Peters (1991) writes that American stock market is characterized by chaotic dynamics. In his second book, Peters (1994) makes the same reasoning for the Dow Jones: the Dow Jones is a long-term memory series ($H = 0.7$, chapter 8) and its underlying process is chaotic (chapter 16)!

These results seem somewhat inconsistent. How can a series be generated both by a long-term memory process and a chaotic process? Peters writes then that on the short run (horizon less than four years), markets are characterized by long-term memory processes, like fractional Gaussian noise. On the long run (horizon more than four years), markets are characterized by deterministic chaotic dynamics. Once again, these affirmations seem inconsistent, particularly concerning the long memory characteristic on the... short run! Moreover, we think that these interpretations are not valid because of the nature (stochastic and deterministic) of the two kinds of processes.

3.3 DETERMINISM AND STOCHASTICITY

Trajectories generated by chaotic processes appear completely random by standard linear time series methods. However, chaotic processes are truly deterministic. Fluctuations are thus endogenous, they are not the result of exogenous shocks. They are generated by the system itself. On the contrary, long-term memory processes, like ARFIMA processes, are stochastic: after an exogenous shock, series will diverge during a more or less long period from its previous time path. For chaotic processes there is no need to introduce exogenous shock in order to produce divergent trajectories.

This dichotomy between stochastic world and deterministic world seems to be crystallized in the relation $D = 2 - H$. Effectively, recall that the dimension D obtained by this relation is the dimension of a stochastic process (e.g., Peitgen, Jürgens et Saupe, 1992, p. 422). As a consequence, if the object under study is the detection of chaos, this relation cannot be used in order to calculate the fractal dimension of the underlying process. In other words:

• Either the underlying process is stochastic: in this case, a link can be established between fractional differencing parameter d and fractal dimension D. Of course, on these conditions, D is not the dimension of a chaotic process since the latter is necessarily deterministic.

• Or the underlying process is deterministic: in this case, the link between long-term memory and fractal dimension is impossible to be established by the relation $D = 2 - H$ or $D = \frac{3}{2} - d$.

Globally, there is no valid relation between long-term memory and chaotic processes. They constitute however two interesting different alternatives for modeling time series.

References

[1] ABRAHAM-FROIS G. and BERREBI E. (1995), *Instabilité, cycles, chaos*, Economica.

[2] BROCK W.A. (1986), "Distinguishing Random and Deterministic Systems: Abridged Version", *Journal of Economic Theory*, Vol. 40, pp. 168–195.

[3] BROCKWELL P.J. and DAVIS R.A. (1991), *Time series: Theory and methods*, Springer Verlag.

[4] ECKMANN J-P. and RUELLE D. (1985), "Ergodic Theory of Chaos and Strange Attractors", *Reviews of Modern Physics*, Vol. 57(3), pp. 617–656.

[5] GEWEKE J. and PORTER-HUDAK S. (1983), "The Estimation and Application of Long Memory Time Series Models", *Journal of Time Series Analysis*, Vol. 4(4), pp.221–238.

[6] GRANGER C.W.J. and JOYEUX R. (1980), "An Introduction To Long-Memory Time Series Models and Fractional Differencing", *Journal of Time Series Analysis*, Vol.1(1),pp.15-29.

[7] GREBOGI C., OTT E., PELIKAN S. and YORKE J.A. (1984), "Strange Attractors that are not Chaotic", *Physica*, 13D, pp. 261–268.

[8] HOSKING J.R.M. (1981), "Fractional Differencing", *Biometrika*, Vol. 68(1).

[9] HURST H.E. (1951), "Long-Term Storage Capacity of Reservoirs", *Transactions of the American Society of Civil Engineers*, Vol. 116, pp. 770–799.

[10] LARDIC S. (1997), "L'hystérèse en économie: définition et mesure", Thèse, UniversitéParisX-Nanterre.

[11] LARDIC S. and MIGNON V. (1996a), "Les tests de mémoire longue appartiennent-ils au camp du démon ?", *Revue Economique*, vol. 47(3), pp. 531–540.

[12] LARDIC S. and MIGNON V. (1996b), "Vingt ans de tests de mémoire longue au travers des processus ARFIMA", Proceedings of the Applied Econometrics Association Symposium, January, Paris.

[13] LARDIC S. and MIGNON V. (1996c), "Essai de mesure du degré de la mémoire longue des séries. L'exemple de la modélisation ARFIMA", UniversitéParis X - Nanterre.

[14] LO A.W. (1991), "Long-Term Memory in Stock Market Prices", *Econometrica*, Vol.59(5), pp.1279-1313.

[15] MANDELBROT B.B. (1972), "Statistical Methodology for Nonperiodic Cycles: From the Covariance to R/S Analysis", *Annals of Economic and Social Measurement*, Vol.1(3),pp.259-290.

[16] MANDELBROT B.B. and VAN NESS J.W. (1968), "Fractional Brownian Motions, Fractional Noises and Applications", *SIAM Review*, Vol. 10(4), pp.422-437.

[17] MANDELBROT B.B. and WALLIS J.R. (1968), "Noah, Joseph and Operational Hydrology", *Water Resources Research*, 4(5), pp. 909–918.

[18] MANDELBROT B.B. and WALLIS J.R. (1969a), "Computer Experiments with Fractional Gaussian Noises", Water Resources Research, 5, pp. 228–267.

[19] MANDELBROT B.B. and WALLIS J.R. (1969b), "Some Long Run Properties of Geophysical Records", Water Resources Research, 5(2), pp. 321–340.

[20] MANDELBROT B.B. and WALLIS J.R. (1969c), "Robustness of the Rescaled Range R/S in the Measurement of Noncyclic Long Run Statistical Dependence", *Water Resources Research*, 5 (5), pp.967-988.

[21] MIGNON V. (1996), "Les implications de la mémoire longue et de la non linéarité sur l'efficience du marché des changes", *Journal de la Société Statistique de Paris*, tome 137 (1), pp.51-72.

[22] PEITGEN H.O., JÜRGENS H. and SAUPE D. (1992), Fractals for the Classroom, Springer Verlag.

[23] PETERS E.E. (1991), Chaos and Order in the Capital Markets, John Wiley & Sons.

[24] PETERS E.E. (1994), Fractal Market Analysis, John Wiley & Sons.

[25] RUELLE D. and TAKENS F. (1971), "On the Nature of Turbulence", *Communications in Mathematical Physics*, 20, pp. 167–192.

[26] SOWELL F. (1992a), "Maximum Likelihood Estimation of Stationary Univariate Fractionally Integrated Time Series Models", *Journal of Econometrics*, Vol. 53, pp. 165-188.

[27] SOWELL F. (1992b), "Modeling Long-Run Behavior with the Fractional ARIMA Model", *Journal of Monetary Economics*, Vol. 29, pp. 277–302.

Contributors

ABRAHAM-FROIS Gilbert, Professor, University of Parix X-Nanterre - UFR SEGMI - MODEM - 200 Avenue de la République - 92001 NANTERRE Cedex, FRANCE.

ARENA Richard, Professor, University of Nice-Sophia Antipolis, Latapses/ CNRS, 250 Rue A. Einstein - 06560 VALBONNE, FRANCE.

BERREBI Edmond, Professor, University of Paris X-Nanterre - UFR SEGMI- MODEM - 200, Avenue de la République - 92001 NANTERRE Cedex, FRANCE.

BOTOMAZAVA Michel, SPPM (Macroeconomic Planning Department), Ministry of Economy and Finance, 101 ANTANANARIVO, MADAGAS-CAR.

DAY H. Richard, Professor, University of Southern California, Department of Economics, 3620 S. Vermont Avenue KAP300 - Los Angeles, CA 90089-0253 USA.

DUFRENOT Gilles, Maître de Conférences, University of Paris XII-Val-de-Marne, ERUDITE, 61, Avenue du Général-de-Gaulle- 94010 CRETEIL Cedex, FRANCE.

FRANKE Reiner, Professor, University of Bremen, Department of Economics, 28334 BREMEN - GERMANY

LARDIC Sendrine, Maître de Conférences, University of Paris X-Nanterre, UFR SEGMI - MODEM, 200, Avenué de la République - 92001 NANTERRE Cedex, FRANCE.

MATHIEU Laurent, Maître de Conférences, University of Paris X-Nanterre, UFR SEGMI - MODEM, 200, Avenue de la République - 92001 NANTERRE Cedex, FRANCE.

MEDIO Alfredo, Professor, University Ca'Foscari - Dipartimento di Scienze Economiche - Dorsoduro 3246 - 30123 VENEZIO - ITALIE.

MIGNON Valérie, Docteur-ATER, University of Paris X-Nanterre - UFR SEGMI-MODEM - 200, Avenue de la République - 92001 NANTERRE Cedex, FRANCE.

RAYBAUT Alain, Chargé de Recherches au CNRS, University of Nice-Sophia Antipolis - Latapses/CNRS - 250 Rue A. Einstein - 06560 VALBONNE, FRANCE.

TOUZÉ Vincent, Ph. D. Candidate, CREST, THEMA and University of Paris X, Timbre J360, 15 boulevard Gabriel Péri, 92245 MALAKOFF, FRANCE.

WANG Zhigang, Ph. D., University of Southern California, Department of Economics, 3620 S. Vermont Avenue, KAP 300, Los Angeles CA 90089-0253 USA.

ZHANG Min, Ph. D. Candidate, University of California, Department of Economics, 3620 S. Vermont Avenue, KAP 300, Los Angeles, CA 90089-0253 USA.